ASE Test Preparation Series

Automobile Test

Manual Drive Trains
and Axles
(Test A3)

4th Edition

THOMSON

DELMAR LEARNING

Australia   Canada   Mexico   Singapore   Spain   United Kingdom   United States

THOMSON
™
DELMAR LEARNING

Thomson Delmar Learning's ASE Test Preparation Series
Automobile Test for Manual Drive Trains and Axles (Test A3), 4th Edition

Vice President, Technology Professional Business Unit:
Gregory L. Clayton

Product Development Manager:
Kristen Davis

Product Manager:
Kim Blakey

Editorial Assistant:
Vanessa Carlson

Director of Marketing:
Beth A. Lutz

Marketing Specialist:
Brian McGrath

Marketing Coordinator:
Marissa Maiella

Production Manager:
Andrew Crouth

Production Editor:
Kara A. DiCaterino

Senior Project Editor:
Christopher Chien

XML Architect:
Jean Kaplansky

Cover Design:
Michael Egan

Cover Images:
Portion courtesy of DaimlerChrysler Corporation

ISBN: 1-4180-3880-6

NOTICE TO THE READER

Publisher does not warrant or guarantee any of the products described herein or perform any independent analysis in connection with any of the product information contained herein. Publisher does not assume, and expressly disclaims, any obligation to obtain and include information other than that provided to it by the manufacturer.

The reader is expressly warned to consider and adopt all safety precautions that might be indicated by the activities herein and to avoid all potential hazards. By following the instructions contained herein, the reader willingly assumes all risks in connection with such instructions.

The publisher makes no representation or warranties of any kind, including but not limited to, the warranties of fitness for particular purpose or merchantability, nor are any such representations implied with respect to the material set forth herein, and the publisher takes no responsibility with respect to such material. The publisher shall not be liable for any special, consequential, or exemplary damages resulting, in whole or part, from the readers' use of, or reliance upon, this material.

Contents

Section 1 The History and Purpose of ASE

Section 2 Take and Pass Every ASE Test

Section 3 Types of Questions on an ASE Exam

Section 4 Overview of the Task List

Section   5   Sample Test for Practice

Section   6   Additional Test Questions for Practice

Section   7   Appendices

Preface

Delmar Learning is very pleased that you have chosen our ASE Test Preparation Series to prepare yourself for the automotive ASE Examination. These guides are available for all of the automotive areas including A1–A8, the L1 Advanced Diagnostic Certification, the P2 Parts Specialist, the C1 Service Consultant and the X1 Undercar Specialist. These guides are designed to introduce you to the Task List for the test you are preparing to take, give you an understanding of what you are expected to be able to do in each task, and take you through sample test questions formatted in the same way the ASE tests are structured.

If you have a basic working knowledge of the discipline you are testing for, you will find Delmar Learning's ASE Test Preparation Series to be an excellent way to understand the "must know" items to pass the test. These books are not textbooks. Their objective is to prepare the technician who has the requisite experience and schooling to challenge ASE testing. It cannot replace the hands-on experience or the theoretical knowledge required by ASE to master vehicle repair technology. If you are unable to understand more than a few of the questions and their explanations in this book, it could be that you require either more shop-floor experience or further study. Some resources that can assist you with further study are listed on the rear cover of this book.

Each book begins with an item-by-item overview of the ASE Task List with explanations of the minimum knowledge you must possess to answer questions related to the task. Following that there are 2 sets of sample questions followed by an answer key to each test and an explanation of the answers to each question. A few of the questions are not strictly ASE format but were included because they help teach a critical concept that will appear on the test. We suggest that you read the complete Task List Overview before taking the first sample test. After taking the first test, score yourself and read the explanation to any questions that you were not sure about, including the questions you answered correctly. Each test question has a reference back to the related task or tasks that it covers. This will help you to go back and read over any area of the task list that you are having trouble with. Once you are satisfied that you have all of your questions answered from the first sample test, take the additional tests and check them. If you pass these tests, you will be prepared to do well on the ASE test.

Our Commitment to Excellence

The 4th edition of Delmar Learning's ASE Test Preparation Series has been through a major revision with extensive updates to the ASE's task lists, test questions, and answers and explanations. Delmar Learning has sought out the best technicians in the country to help with the updating and revision of each of the books in the series.

About the Series Advisor

To promote consistency throughout the series, a series advisor took on the task of reading, editing, and helping each of our experts give each book the highest level of accuracy possible. Dan Perrin has served in the role of Series Advisor for the 4th edition of the ASE Test Preparation Series. Dan began ASE testing with the first series of tests in 1972 and has been continually certified ever since. He holds ASE master status in automotive, truck, collision, and machinist. He is also L1, L2, and alternated fuels certified, along with some others that have expired. He has been an automotive educator since 1979, having taught at the secondary, post-secondary, and industry levels. His service includes participation on boards that include the North American Council of Automotive Teachers (NACAT), the Automotive Industry Planning Council (AIPC), and the National Automotive Technicians Education Foundation (NATEF). Dan currently serves as the Executive Manager of NACAT and Director of the NACAT Education Foundation.

Thanks for choosing Delmar Learning's ASE Test Preparation Series. All of the writers, editors, Delmar Staff, and myself have worked very hard to make this series second to none. I know you are going to find this book accurate and easy to work with. It is our objective to constantly improve our product at Delmar by responding to feedback.

If you have any questions concerning the books in this series, you can email me at: autoexpert@trainingbay.com.

Dan Perrin
Series Advisor

1 The History and Purpose of ASE

ASE began as the National Institute for Automotive Service Excellence (NIASE). It was founded as a non-profit independent entity in 1972 by a group of industry leaders with the single goal of providing a means for consumers to distinguish between incompetent and competent technicians. It accomplishes this goal by testing and certification of repair and service professionals. From this beginning it has evolved to be known simply as ASE (Automotive Service Excellence) and today offers more than 40 certification exams in automotive, medium/heavy duty truck, collision, engine machinist, school bus, parts specialist, automobile service consultant, and other industry-related areas. At this time there are more than 400,000 professionals with current ASE certifications. These professionals are employed by new car and truck dealerships, independent garages, fleets, service stations, franchised service facilities, and more. ASE continues its mission by also providing information that helps consumers identify repair facilities that employ certified professionals through its Blue Seal of Excellence Recognition Program. Shops that have a minimum of 75% of their repair technicians ASE certified and meet other criteria can apply for and receive the Blue Seal of Excellence Recognition from ASE.

ASE recognized that educational programs serving the service and repair industry also needed a way to be recognized as having the faculty, facilities, and equipment to provide a quality education to students wanting to become service professionals. Through the combined efforts of ASE, industry, and education leaders, the non-profit National Automotive Technicians Education Foundation (NATEF) was created to evaluate and recognize training programs. Today more than 2000 programs are ASE certified under the standards set by the service industry. ASE/NATEF also has a certification of industry (factory) training program known as CASE. CASE stands for Continuing Automotive Service Education and recognizes training provided by replacement parts manufacturers as well as vehicle manufacturers.

ASE certification testing is administered by the American College Testing (ACT). Strict standards of security and supervision at the test centers insure that the technician who holds the certification earned it. Additionally ASE certification also requires that the person passing the test to be able to demonstrate that they have two years of work experience in the field before they can be certified. Test questions are developed by industry experts that are actually working in the field being tested. There is more detail on how the test is developed and administered in the next section. Paper and pencil tests are administered twice a year at over seven hundred locations in the United States. Computer based testing is now also available with the benefit of instant test results at certain established test centers. The certification is valid for five years and can be recertified by retesting. So that consumers can recognize certified technicians, ASE issues a jacket patch, certificate, and wallet card to certified technicians and makes signs available to facilities that employ ASE certified technicians.

You can contact ASE at any of the following:

National Institute for Automotive Service Excellence
101 Blue Seal Drive S.E.
Suite 101
Leesburg, VA 20175
Telephone 703-669-6600
FAX 703-669-6123
www.ase.com

WE SUPPORT
PROFESSIONAL CERTIFICATION
THROUGH THE
National Institute for
AUTOMOTIVE
SERVICE
EXCELLENCE

2 Take and Pass Every ASE Test

Participating in an Automotive Service Excellence (ASE) voluntary certification program gives you a chance to show your customers that you have the "know-how" needed to work on today's modern vehicles. The ASE certification tests allow you to compare your skills and knowledge to the automotive service industry's standards for each specialty area.

If you are the "average" automotive technician taking this test, you are in your mid-thirties and have not attended school for about fifteen years. That means you probably have not taken a test in many years. Some of you, on the other hand, have attended college or taken postsecondary education courses and may be more familiar with taking tests and with test-taking strategies. There is, however, a difference in the ASE test you are preparing to take and the educational tests you may be accustomed to.

How are the tests administered?

ASE test are administered at over 750 test sites in local communities. Paper and pencil tests are the type most widely available to technicians. Each tester is given a booklet containing questions with charts and diagrams where required. You can mark in this test booklet but no information entered in the booklet is scored. Answers are recorded on a separate answer sheet. You will enter your answers, using a number 2 pencil only. ASE recommends you bring four sharpened number 2 pencils that have erasers. Answer choices are recorded by coloring in the blocks on the answer sheet. The answer sheets are scanned electronically and the answers tabulated. For test security, test booklets include randomly generated questions. Your answer key must be matched to the proper booklet so it is important to correctly enter the booklet serial number on the answer sheet. All instructions are printed on the test materials and should be followed carefully.

ASE has introduced Computer Based Testing (CBT) at some locations. While the test content is the same for both testing methods the CBT tests have some unique requirements and advantages. It is strongly recommended that technicians considering the CBT tests go the ASE web page at www.ASE.com and review the conditions and requirements for this type of test. There is a demonstration of a CBT that allows you to experience this type of test before you register. Some technicians find this style of testing provides an advantage, while others find operating the computer a distraction. One significant benefit of CBT is the availability of instant results. You can receive your test results before you leave the test center. CBT testing also offers increased flexibility in scheduling. The cost for taking CBTs is slightly higher than paper and pencil tests and the number of testing sites is limited. The first time test taker may be more comfortable with the paper and pencil tests but technicians now have a choice.

Who Writes the Questions?

The questions are written by service industry experts in the area being tested. Each area will have its own technical experts. Questions are entirely job related. They are designed to test the skills you need to be a successful technician. Theoretical knowledge is important and necessary to answer the questions, but the ability to apply that knowledge is the basis of ASE test questions.

Each question has its roots in an ASE "item-writing" workshop where service representatives from automobile manufacturers (domestic and import), aftermarket parts and equipment manufacturers, working technicians, and vocational educators meet in a workshop setting to share ideas and translate them into test questions. Each test question written by these experts must survive review by all members of the group.

The questions are written to deal with practical application of soft skills and system knowledge experienced by technicians in their day-to-day work.

All questions are pre-tested and quality-checked on a national sample of technicians. Those questions that meet ASE standards of quality and accuracy are included in the scored sections of the tests; the "rejects" are sent back to the drawing board or discarded altogether.

Each certification test is made up of between forty and eighty multiple-choice questions.

Note: Each test could contain additional questions that are included for statistical research purposes only. Your answers to these questions will not affect your score, but since you do not know which ones they are, you should answer all questions on the test. The five-year Recertification Test will cover the same content areas as those listed above. However, the number of questions in each content area of the Recertification Test will be reduced by about one-half.

Objective Tests

A test is called an objective test if the same standards and conditions apply to everyone taking the test and there is only one correct answer to each question.

Objective tests primarily measure your ability to recall information. A well-designed objective test can also test your ability to understand, analyze, interpret, and apply your knowledge. Objective tests include true-false, multiple choice, fill in the blank, and matching questions. ASE's tests consist exclusively of four-part multiple-choice objective questions.

The following are some strategies that may be applied to your tests.

Before beginning to take an objective test, quickly look over the test to determine the number of questions, but do not try to read through all of the questions. In an ASE test, there are usually between forty and eighty questions, depending on the subject. Read through each question before marking your answer. Answer the questions in the order they appear on the test. Leave the questions blank that you are not sure of and move on to the next question. You can return to those unanswered questions after you have finished the others. They may be easier to answer at a later time after your mind has had additional time to consider them on a subconscious level. In addition, you might find information in other questions that will help you recall the answers to some of them.

Do not be obsessed by the apparent pattern of responses. For example, do not be influenced by a pattern like **D, C, B, A, D, C, B, A** on an ASE test.

There is also a lot of folk wisdom about taking objective tests. For example, there are those who would advise you to avoid response options that use certain words such as *all, none, always, never, must,* and *only,* to name a few. This, they claim, is because nothing in life is exclusive. They would advise you to choose response options that use words that allow for some exception, such as *sometimes, frequently, rarely, often, usually, seldom,* and *normally.* They would also advise you to avoid the first and last option (A and D) because test writers, they feel, are more comfortable if they put the correct answer in the middle (B and C) of the choices. Another recommendation often offered is to select the option that is either shorter or longer than the other three choices because it is more likely to be correct. Some would advise you to never change an answer since your first intuition is usually correct.

Although there may be a grain of truth in this folk wisdom, ASE test writers try to avoid them and so should you. There are just as many **A** answers as there are **B** answers, just as many **D** answers as **C** answers. As a matter of fact, ASE tries to balance the answers at about 25 percent per choice **A, B, C,** and **D.** There is no intention to use "tricky" words, such as outlined above. Put no credence in the opposing words "sometimes" and "never," for example.

Multiple-choice tests are sometimes challenging because there are often several choices that may seem possible, and it may be difficult to decide on the correct choice. The best strategy, in this case, is to first determine the correct answer before looking at the options. If you see the answer you decided on, you should still examine the options to make sure that none seem more correct than yours. If you do not know or are not sure of the answer, read each option very carefully and try to eliminate those options that you know to be wrong. That way, you can often arrive at the correct choice through a process of elimination.

If you have gone through all of the test and you still do not know the answer to some of the questions, then guess. Yes, guess. You then have at least a 25 percent chance of being correct. If you leave the question blank, you have no chance. Your score is based on the number of questions answered correctly.

Preparing for the Exam

The main reason we have included so many sample and practice questions in this guide is, simply, to help you learn what you know and what you don't know. We recommend that you work your way through each question in this book. Before doing this, carefully look through Section 3; it contains a description and explanation of the question types you'll find on an ASE exam.

Once you understand what the questions will look like, move to the sample test. Answer one of the sample questions (Section 5) then read the explanation (Section 7) to the answer for that question. If you don't feel you understand the reasoning for the correct answer, go back and read the overview (Section 4) for the task that is related to that question. If you still don't feel you have a solid understanding of the material, identify a good source of information on the topic, such as a textbook, and do some more studying.

After you have completed all of the sample test items and reviewed your answers, move to the additional questions (Section 6). This time answer the questions as if you were taking an actual test. Do not use any reference or allow any interruptions in order to get a feel for how you will do on an actual test. Once you have answered all of the questions, grade your results using the answer key in Section 7. For every question that you gave a wrong answer to, study the explanations to the answers and/or the overview of the related task areas. Try to determine the root cause for your missing the question. The easiest thing to correct is learning the correct technical content. The hardest thing to correct is behaviors that lead you to a wrong conclusion. If you knew the information but still got it wrong there is a behavior problem that will need to be corrected. An example would be reading too quickly and skipping over words that affect your reasoning. If you can identify what you did that caused you to answer the question incorrectly you can eliminate that cause and improve your score. Here are some basic guidelines to follow while preparing for the exam:

- Focus your studies on those areas you are weak in.

- Be honest with yourself while determining if you understand something.

- Study often but in short periods of time.

- Remove yourself from all distractions while studying.

- Keep in mind the goal of studying is not just to pass the exam, the real goal is to learn!

- Prepare physically by getting a good night's rest before the test and eat meals that provide energy but do not cause discomfort.

- Arrive early to the test site to avoid long waits as test candidates check in and to allow all of the time available for your tests.

During the Test

On paper and pencil tests you will be placing your answers on a sheet where you will be required to color in your answer choice. Stray marks or incomplete erasures may be picked up as an answer by the electronic reader, so be sure only your answers end up on the sheet. One of the biggest problems an adult faces in test taking, it seems, is placing the answer in the correct spot on the answer sheet. Make certain that you mark your answer for, say, question 21, in the space on the answer sheet designated for the answer for question 21. A correct response in the wrong line will probably result in two questions being marked wrong, one with two answers (which could include a correct answer but will be scored wrong) and the other with no answer. Remember, the answer sheet on the written test is machine scored and can only "read" what you have colored in.

If you finish answering all of the questions on a test and have remaining time, go back and review the answers to those questions that you were not sure of. You can often catch careless errors by using the remaining time to review your answers. Carefully check your answer sheet for blank answer blocks or missing information.

At practically every test, some technicians will invariably finish ahead of time and turn their papers in long before the final call. Some technicians may be doing recertification tests and others may be taking fewer tests than you. Do not let them distract or intimidate you.

It is not wise to use less than the total amount of time that you are allotted for a test. If there are any doubts, take the time for review. Any product can usually be made better with some additional effort. A test is no exception. It is not necessary to turn in your test paper until you are told to do so.

Testing Time Length

An ASE written test session is four hours. You may attempt from one to a maximum of four tests in one session. It is recommended, however, that no more than a total of 225 questions be attempted at any test session. This will allow for just over one minute for each question.

Visitors are not permitted at any time. If you wish to leave the test room, for any reason, you must first ask permission. If you finish your test early and wish to leave, you are permitted to do so only during specified dismissal periods.

You should monitor your progress and set an arbitrary limit to how much time you will need for each question. This should be based on the number of questions you are attempting. It is suggested that you wear a watch because some facilities may not have a clock visible to all areas of the room.

Computer-Based Tests are allotted a testing time according to the number of questions ranging from one half hour to one and one half hours. Advanced level tests are allowed two hours. This time is by appointment and you should be sure to be on time to insure that you have all of the time allocated. If you arrive late for a CBT test appointment you will only have the amount of time remaining on your appointment.

Your Test Results!

You can gain a better perspective about tests if you know and understand how they are scored. ASE's tests are scored by American College Testing (ACT), a nonpartial, unbiased organization having no vested interest in ASE or in the automotive industry.

Each question carries the same weight as any other question. For example, if there are fifty questions, each is worth 2 percent of the total score. The passing grade is 70 percent. That means you must correctly answer thirty-five of the fifty questions to pass the test.

The test results can tell you:

- where your knowledge equals or exceeds that needed for competent performance, or

- where you might need more preparation.

Your ASE test score report is divided into content areas and will show the number of questions in each content area and how many of your answers were correct. These numbers provide information about your performance in each area of the test. However, because there may be a different number of questions in each content area of the test, a high percentage of correct answers in an area with few questions may not offset a low percentage in an area with many questions.

It should be noted that one does not "fail" an ASE test. The technician who does not pass is simply told "More Preparation Needed." Though large differences in percentages may indicate problem areas, it is important to consider how many questions were asked in each area. Since each test evaluates all phases of the work involved in a service specialty, you should be prepared in each area. A low score in one area could keep you from passing an entire test.

There is no such thing as average. You cannot determine your overall test score by adding the percentages given for each task area and dividing by the number of areas. It doesn't work that way because there generally are not the same number of questions in each task area. A task area with twenty questions, for example, counts more toward your total score than a task area with ten questions.

Your test report should give you a good picture of your results and a better understanding of your strengths and weaknesses for each task area.

If you fail to pass the test, you may take it again at any time it is scheduled to be administered. You are the only one who will receive your test score. Test scores will not be given over the telephone by ASE nor will they be released to anyone without your written permission.

3 Types of Questions on an ASE Exam

ASE certification tests are often thought of as being tricky. They may seem to be tricky if you do not completely understand what is being asked. The following examples will help you recognize certain types of ASE questions and avoid common errors.

Paper-and-pencil tests and computer-based test questions are identical in content and difficulty. Most initial certification tests are made up of forty to eighty multiple-choice questions. Multiple-choice questions are an efficient way to test knowledge. To answer them correctly, you must think about each choice as a possibility, and then choose the one that best answers the question. To do this, read each word of the question carefully. Do not assume you know what the question is about until you have finished reading it.

About 10 percent of the questions on an actual ASE exam will use an illustration. These drawings contain the information needed to correctly answer the question. The illustration must be studied carefully before attempting to answer the question. Often, technicians look at the possible answers then try to match up the answers with the drawing. Always do the opposite; match the drawing to the answers. When the illustration is showing an electrical schematic or another system in detail, look over the system and try to figure out how the system works before you look at the question and the possible answers.

Multiple-Choice Questions

The most common type of question used on ASE Tests is the multiple-choice question. This type of question contains three "distracters" (wrong answers) and one "key" (correct answer). When the questions are written effort is made to make the distracters plausible to draw an inexperienced technician to one of them. This type of question gives a clear indication of the technician's knowledge. Using multiple criteria including cross-sections by age, race, and other background information, ASE is able to guarantee that a question does not bias for or against any particular group. A question that shows bias toward any particular group is discarded. If you encounter a question that you are unsure of, reverse engineer it by eliminating the items that it cannot be. For example:

A rocker panel is a structural member of which vehicle construction type?

A. Front-wheel drive
B. Pickup truck
C. Unibody
D. Full-frame

Analysis:

This question asks for a specific answer. By carefully reading the question, you will find that it asks for a construction type that uses the rocker panel as a structural part of the vehicle.

Answer A is wrong. Front-wheel drive is not a vehicle construction type.
Answer B is wrong. A pickup truck is not a type of vehicle construction.
Answer C is correct. Unibody design creates structural integrity

by welding parts together, such as the rocker panels, but does not require exterior cosmetic panels installed for full strength.

Answer D is wrong. Full-frame describes a body-over-frame construction type that relies on the frame assembly for structural integrity.

Therefore, the correct answer is C. If the question was read quickly and the words "construction type" were passed over, answer A may have been selected.

EXCEPT Questions

Another type of question used on ASE tests has answers that are all correct except one. The correct answer for this type of question is the answer that is wrong. The word "**EXCEPT**" will always be in capital letters. You must identify which of the choices is the wrong answer. If you read quickly through the question, you may overlook what the question is asking and answer the question with the first correct statement. This will make your answer wrong. An example of this type of question and the analysis is as follows:

All of the following are tools for the analysis of structural damage **EXCEPT**:

A. height gauge
B. tape measure.
C. dial indicator.
D. tram gauge.

Analysis:

The question really requires you to identify the tool that is not used for analyzing structural damage. All tools given in the choices are used for analyzing structural damage except one. This question presents two basic problems for the test-taker who reads through the question too quickly. It may be possible to read over the word "**EXCEPT**" in the question or not think about which type of damage analysis would use answer C. In either case, the correct answer may not be selected. To correctly answer this question, you should know what tools are used for the analysis of structural damage. If you cannot immediately recognize the incorrect tool, you should be able to identify it by analyzing the other choices.

Answer A is wrong. A height gauge may be used to analyze structural damage.
Answer B is wrong. A tape measure may be used to analyze structural damage.
Answer C is correct. A dial indicator may be used as a damage analysis tool for moving parts, such as wheels, wheel hubs, and axle shafts, but would not be used to measure structural damage.
Answer D is wrong. A tram gauge is used to measure structural damage.

Technician A, Technician B Questions

The type of question that is most popularly associated with an ASE test is the "Technician A says . . . Technician B says . . . Who is right?" type. In this type of question, you must identify the correct statement or statements. To answer this type of question correctly, you must carefully read each technician's statement and judge it on its own merit to determine if the statement is true.

Sometimes this type of question begins with a statement about some analysis or repair procedure. This is often referred to as the stem of the question and provides the setup or background information required to understand the conditions the question is based on. This is followed by two statements about the cause of the concern, proper inspection, identification, or repair choices. You are asked whether the first statement, the second statement, both statements, or neither statement is correct. Analyzing this type of question is a little easier than the other types because there are only two ideas to consider although there are still four choices for an answer.

Technician A, Technician B questions are really double true or false questions. The best way to analyze this kind of question is to consider each technician's statement separately. Ask yourself, is A true or false? Is B true or false? Then select your answer from the four choices. An important point to remember is that an ASE Technician A, Technician B question will never have Technician A and B directly disagreeing with each other. That is why you must evaluate each statement independently.

An example of this type of question and the analysis of it follows.

A vehicle comes into the shop with a gas gauge that will not register above one half full. When the sending unit circuit is disconnected the gauge reads empty and when it is connected to ground the gauge goes to full. Technician A says that the sending unit is shorted to ground. Technician B says the gauge circuit is working and the sending unit is likely the problem. Who is right?

A. A only
B. B only
C. Both A and B
D. Neither A nor B

Analysis:

Reading of the stem of the question sets the conditions of the customer concern and establishes what information is gained from testing. General knowledge of gauge circuits and test procedures are needed to correctly evaluate the technician's conclusions. Note: Avoid being distracted by experience with unusual or problem vehicles that you may have worked on, Other technicians taking the same test do not have that knowledge, so it should not be used as the basis of your answers.

Technician A is wrong because a shorted to ground sending unit would produce a gauge reading equivalent to the test conditions of a grounding the circuit and produce a full reading. **Technician B is correct** because the gauge spans when going from an open circuit to a completely
grounded circuit. This would tend to indicate that the problem had to be in the sending unit.
Answer C is not correct. Both technicians are identifying the problem as a sending unit but technician A qualified the problem as a specific type of failure (grounded) that would not have caused the symptoms of the vehicle.
Answer D is not correct because technician B's diagnosis is a possible cause of the conditions identified.

Most-Likely Questions

Most-Likely questions are somewhat difficult because only one choice is correct while the other three choices are nearly correct. An example of a Most-Likely-cause question is as follows:

The Most-Likely cause of reduced turbocharger boost pressure may be a:

A. wastegate valve stuck closed.
B. wastegate valve stuck open.
C. leaking wastegate diaphragm.
D. disconnected wastegate linkage.

Analysis:

Answer A is wrong. A wastegate valve stuck closed increases turbocharger boost pressure.
Answer B is correct. A wastegate valve stuck open decreases turbocharger boost pressure.
Answer C is wrong. A leaking wastegate valve diaphragm increases turbocharger boost pressure.

Answer D is wrong. A disconnected wastegate valve linkage will increase turbocharger boost pressure.

LEAST-Likely Questions

Notice that in Most-Likely questions there is no capitalization. This is not so with LEAST-Likely type questions. For this type of question, look for the choice that would be the LEAST-Likely cause of the described situation. Read the entire question carefully before choosing your answer. An example is as follows:

What is the LEAST-Likely cause of a bent pushrod?

A. Excessive engine speed
B. A sticking valve
C. Excessive valve guide clearance
D. A worn rocker arm stud

Analysis:

Answer A is wrong. Excessive engine speed may cause a bent pushrod.
Answer B is wrong. A sticking valve may cause a bent pushrod.
Answer C is correct. Excessive valve clearance will not generally cause a bent pushrod.
Answer D is wrong. A worn rocker arm stud may cause a bent pushrod.

You should avoid relating questions to those unusual situations that you may have encountered and answer based on the technical and mechanical possibilities.

Summary

There are no four-part multiple-choice ASE questions having "none of the above" or "all of the above" choices. ASE does not use other types of questions, such as fill-in-the-blank, completion, true-false, word-matching, or essay. ASE does not require you to draw diagrams or sketches. If a formula or chart is required to answer a question, it is provided for you. There are no ASE questions that require you to use a pocket calculator.

4 | Overview of the Task List

Manual Drive Trains and Axles (Test A3)

The following section includes the task areas and task lists for this test and a written overview of the topics covered in the test.

The task list describes the actual work you should be able to do as a technician that you will be tested on by the ASE. This is your key to the test and you should review this section carefully. We have based our sample test and additional questions upon these tasks. The overview section will also support your understanding of the task list. ASE advises that the questions on the test may not equal the number of tasks listed; the task lists tell you what ASE expects you to know how to do and be ready to be tested upon.

At the end of each question in the Sample Test and Additional Test Questions sections, a letter and number will be used as a reference back to this section for additional study. Note the following example: **B.11.**

B. Transmission Diagnosis and Repair (7 Questions)

Task B.11   **Inspect and replace reverse idler gear, shaft, bearings, thrust washers, and retainers/snap rings.**

Example:

1. While inspecting a reverse idler gear from a manual transmission, Technician A says that you should check the center bore for a smooth mar-free surface. Technician B says that the reverse idler gear is splined in the center bore and that the splines should be checked for excessive wear or damage. Who is right?
    A. Technician A only
    B. Technician B only
    C. Both A and B
    D. Neither A nor B.   (B.11)

Analysis:

Question #1
Answer A is correct. Only Technician A is correct. The center bore of the idler gear should be smooth and mar free.
Answer B is wrong. The reverse idler gear rides on needle bearings and is not splined in the center bore.
Answer C is wrong. Only Technician A is correct.
Answer D is wrong. Only Technician A is correct. The reverse idler gear teeth should be inspected for chips, pits, and cracks. Check the gear bore for roughness and scoring. The needle bearings/bushings and shafts must be inspected for roughness, scoring, and pitting.

Task List and Overview

A. Clutch Diagnosis and Repair (6 Questions)

Task A.1   **Diagnose clutch noise, binding, slippage, pulsation, chatter, pedal feel/effort, and release problems; determine needed repairs.**

Clutch chatter is felt normally when the pressure plate has made initial contact with the flywheel. Things that can cause a clutch to chatter, such as weak torsional springs, will not absorb the shock of the clutch when contact is made. Other causes of chatter could be a bent clutch disc or a burned or glazed lining on the disc. If the flywheel is glazed, has excessive runout, or is scored, clutch chatter may occur.

If a clutch was disengaged, the input shaft on the transmission would not be spinning, eliminating the bearing noise coming from the input shaft bearing. A pilot bearing would be a probable cause of noise in this case. With the input shaft not spinning, the pilot bearing is spinning on the end of the input shaft. This could cause a bearing noise due to a bad or worn pilot bearing.

Task A.2   **Inspect, adjust, and replace clutch pedal linkage, cables and automatic adjuster mechanisms, brackets, bushings, pivots, springs, and electrical switches.**

When there is no clutch pedal free play, the clutch is not fully engaged. The release bearing is touching the fingers of the pressure plate. This will relieve some of the clamping on the clutch disc. This may cause the clutch to slip; it will not cause hard shifting, improper clutch release, or transaxle gear damage. These would all be signs of too much clutch pedal free play.

Many late-model vehicles have self-adjusting cables. The cable is adjusted when the clutch pedal is released as the clutch disc wears from normal use. These systems use a constant-running release bearing. It is always in contact with the pressure plate. The clutch pedal will not have any clutch pedal free play.

Task A.3   **Inspect, adjust, replace, and bleed hydraulic clutch slave cylinder/actuator, master cylinder, lines, and hoses; clean and flush hydraulic system; refill with proper fluid.**

The hydraulic system for a hydraulic clutch is totally separate from the brake system. The clutch master cylinder is mounted to the fire wall in the engine compartment. A line runs from the clutch master cylinder to the slave cylinder on the bell housing, or inside the bell housing on some models. A rod connected to the clutch pedal goes through the fire wall into the clutch master cylinder. When the rod is pushed into the clutch master cylinder, it forces fluid through the line, which actuates the slave cylinder to release the clutch.

Air in the system will prevent the clutch from disengaging properly when the clutch pedal is fully depressed. Air is compressible and will compress in the hydraulic system before the spring pressure in the pressure plate releases the clutch disc from the flywheel. The conditions of less free play, worn facings, or a scored pressure plate will cause clutch problems, but will not affect the release of the clutch.

Task A.4   **Inspect, adjust, and replace release (throw-out) bearing, bearing retainer, lever, and pivot.**

On a clutch with an adjustable linkage, the release bearing should not be in contact with the pressure plate fingers. If the release bearing is not touching the fingers, it will not make any noise even if the bearing is bad. Clutch pedal free play is the distance between the release bearing and the pressure plate fingers. It is the gap or movement in the clutch pedal before the release bearing contacts the pressure plate and releases the clutch.

Hydraulic-controlled clutch systems use a release bearing that is always in contact with the fingers on the pressure plate. There is no manual adjustment on a hydraulic clutch system; it adjusts automatically as the clutch disc wears. When a clutch is disengaged, the release bearing moves toward the pressure plate. The release bearing continues to move toward the pressure plate fingers and compresses the springs in the pressure plate to release the clutch.

Task A.5   Inspect and replace clutch disc and pressure plate assembly; inspect input shaft splines.

Excessive crankshaft end play causes the pressure plate to move away from the clutch release bearing, which may result in improper clutch release. Loose or worn engine main bearings may cause an oil leak at the rear main bearing seal, which contaminates the clutch facings with oil, resulting in clutch slippage. An improper pressure plate-to-flywheel position causes engine vibrations.

If the clutch facing is worn too thin, the clamping force of the pressure plate will not be as much as it was when the clutch disc was at full thickness. There will not be enough spring pressure left in the pressure plate to maintain a hard clamping force on the clutch disc because the springs in the pressure plate are fully extended and not applying enough pressure. The minimum thickness of the clutch disc lining is 0.012 inch (0.3 mm). Slippage will occur if the lining is any thinner.

Proper installation of the clutch disc is critical; the damper spring offset must face the transmission. The clutch disc is normally marked to indicate which side should face the flywheel. If the damper spring offset is toward the engine, the springs may contact the flywheel or the flywheel bolts, damaging these components.

Task A.6   Inspect and replace pilot bearing/bushing; inspect pilot bearing/bushing mating surfaces.

When the clutch is engaged, the transmission input shaft rotates at the same speed as the engine flywheel and pilot bearing at all times. When the clutch is disengaged, the flywheel and pilot bearing rotate on the end of the transmission input shaft and turn faster than the shaft.

If a bushing-type pilot bearing is lubricated with bearing grease, friction actually will increase between the bushing and the transmission input shaft. Lubricate a bushing-type pilot bearing with motor oil. Lubricate a roller-type pilot bearing with wheel bearing grease.

Task A.7   Inspect and measure flywheel and ring gear; inspect dual-mass flywheel damper where required; repair or replace as necessary.

If too much material is removed from the flywheel, the torsion springs on the clutch plate are moved closer to the mounting bolts on the flywheel, and these springs may contact the heads of the flywheel bolts. Removing excessive material from the flywheel moves the pressure plate forward, away from the release bearing. This action increases free play so the slave cylinder rod may not move far enough to release the clutch.

By not resurfacing the flywheel, damage or premature wear can be caused to the rest of the new assembly that was installed. A flywheel should be resurfaced every time a clutch assembly is replaced. Resurfacing the flywheel will ensure that it has the flatness and a microfinish it needs to ensure that the new clutch disc breaks in properly. If it is not resurfaced, the clutch disc will probably glaze and chatter. The flywheel should be cleaned with hot water and soap to remove all residue on the surface after it has been resurfaced.

Inspect the shaft tip that rides in the pilot bearing for smoothness. Check the splines of the shaft for any wear that could prevent the clutch disc from sliding evenly and smoothly. If the splines have excessive wear or damage, this will cause the clutch to engage roughly. A clutch disc that is bent or that has weak torsional springs will also cause the clutch to engage roughly.

Designs of dual-mass flywheels vary so it is most important to know what they do. A dual-mass flywheel is designed to isolate torsional crankshaft spikes created by diesel engines with high compression ratios. By separating the mass of the flywheel between the engine and the transmission, torsional spikes can be isolated, eliminating potential damage to the transmission gear teeth. The technician will need to inspect the springs or other dampening method used to connect the two separate components of the flywheel.

Task A.8   Inspect engine block, clutch (bell) housing, transmission case mating surfaces, and alignment dowels; determine needed repairs.

If the bell housing is not aligned properly with the engine block because of something being pinched between them or because of a burr or imperfection on one of the mating surfaces, worn alignment dowels, or loose bell housing bolts, the clutch will not make even contact with the flywheel. This will cause the clutch to chatter and grab because of the uneven contact when the clutch is being engaged. Reduced clutch pedal free play, growling noises, or vibrations at high speeds are not symptoms of a misalignment condition.

If a clutch disc or an input shaft is bent due to careless removal or installation of the transmission, this will cause misalignment problems. If the transmission is misaligned when it is installed, this can cause the pilot bushing or bearing to wear out prematurely.

Task A.9   Measure flywheel-to-block runout and crankshaft end play; determine needed repairs.

To measure crankshaft end play, mount the magnetic dial indicator on the back of the engine block. Position the dial indicator to the flywheel. Push the flywheel toward the front of the engine until it stops. Adjust the dial indicator to zero and then pull the flywheel toward the back of the engine. The reading on the dial indicator will be the crankshaft end play.

The flywheel can be checked with the dial indicator setup after checking the crankshaft end play. Observe the dial indicator as the flywheel is rotated. The measured movement on the dial indicator is the runout and should be compared to the vehicle specifications.

Excessive main bearing wear will cause low oil pressure or a rear main oil leak, possibly causing the clutch disc to become contaminated with engine oil. A thrust bearing is placed between the crankshaft main bearing cap and the side of a crankshaft journal. These thrust bearings are put into place to control the forward and rearward movement of the crankshaft during acceleration and deceleration. The proper thickness thrust bearing is selected when the engine is assembled to set the crankshaft end play.

Task A.10   Inspect, replace, and align powertrain mounts.

Engine and transmission mounts should be inspected for broken, sagged, oil-soaked, or deteriorated conditions. Any of these mount conditions may cause a grabbing, binding clutch. On a rear-wheel drive car, damaged engine or transmission mounts may cause improper driveshaft angles, which result in a vibration that changes in intensity when the vehicle accelerates and decelerates.

On a vehicle with a manual clutch, part of the clutch linkage is connected from the frame of the vehicle to the engine block. If the vehicle has a broken engine mount, the engine can lift off the frame on acceleration. This can cause the clutch linkage to move and apply release pressure to the release bearing fork. This may cause the release bearing to apply release pressure on the pressure plate, causing the clutch disc to slip between the flywheel and the pressure plate.

B. Transmission Diagnosis and Repair (7 Questions)

Task B.1   Diagnose transmission noise, hard shifting, gear clash, jumping out of gear, fluid condition, and fluid leakage problems; determine needed repairs.

A misaligned or loose transmission case or clutch housing may cause the transmission to slip out of gear. Check the mounting of the transmission to the engine block for looseness or dirt between the two cases. Broken or loose engine mounts will also cause misalignment. This problem may also be caused by insufficient spring tension of the shift rail detent spring or bent or worn shift forks, levers, or shafts. Other internal problems may cause this condition, such as a worn input shaft pilot bearing, bent output shaft, or a worn or broken synchronizer.

If the transmission shifts hard or the gears clash while shifting, a common cause of the problem is the clutch. Check the clutch pedal free-travel adjustment. Make sure the clutch releases completely. Also check for worn clutch parts and a binding input shaft pilot bearing. Shift linkage problems can also cause this problem. If the shift lever is worn, binding, or out of adjustment, proper engagement of gears is impossible. An unlubricated linkage will also cause shifting difficulties.

Task B.2 Inspect, adjust, and replace transmission external shifter assembly, shift linkages, brackets, bushings/grommets, pivots, and levers.

Most external shift linkages and cables require adjusting, and a similar adjustment procedure is used on some vehicles. Raise the vehicle and place the shifter in the neutral position to begin the shift linkage adjustment. With a lever-type shift linkage, install a rod in the adjustment hole in the shifter assembly. Adjust the shift linkages by loosening the rod retaining locknuts and moving the levers until the rod fully enters the alignment holes. Tighten the locknuts and check the shift operation in all gears.

Transmissions with internal linkage do not have provisions for adjustment. However, external linkages, both floor and column mount, can be adjusted. Linkages are adjusted at the factory, but worn parts may make adjustment necessary. Also, after a transmission has been disassembled, the shift lever may need adjustment.

Task B.3 Inspect and replace transmission gaskets, sealants, seals, and fasteners; inspect sealing surfaces.

Excessive output shaft or excessive input shaft end play results in lateral shaft movement that may adversely affect the extension housing seal. A worn output shaft bearing will not cause premature extension housing seal failure. If the driveshaft yoke has a score or imperfection on the shaft, it could damage the seal and cause the transmission to leak fluid at the rear of the transmission. Replacing the seal will not correct this condition until the yoke is replaced.

A cork gasket should be installed as it is when it comes out of the box. It was made to be installed dry and does not require any type of added sealant to help the gasket seal any better. A spray adhesive sometimes may be used to hold the gasket in place to help installation. Do not use any additional gasket sealant when installing a rubber gasket. It will become too slippery and may not position correctly.

Task B.4 Remove and replace transmission; inspect transmission mounts.

The driveshaft should be installed in the original position on the rear axle pinion flange. If it is not installed in the original position, you may experience a driveline vibration. If the transmission is supported from the input shaft, the weight of the transmission itself could cause the input shaft and the clutch disc to be bent or damaged. The engine-support fixture must be installed before the transmission-to-engine bolts are even loosened. When the transmission is being removed, the clutch disc may move, causing it to be misaligned. A clutch disc alignment tool must be used before the transmission is installed to align the clutch disc with the flywheel.

A transmission mount absorbs a lot of torque and vibration in the rubber mount when shifting and accelerating. If the rubber mount becomes saturated with oil, the oil will deteriorate the rubber and weaken it. This will eventually cause the mount to fail.

Oil will not cause the mount to crack, but the oil will make the rubber in the mount feel soft and spongy.

Task B.5 Disassemble and clean transmission components; reassemble transmission.

A synchronizer hub that does not slide smoothly over the blocker ring causes the hub to jam, resulting in hard shifting. The stronger hub and sleeve should be marked before disassembly so it can be installed the same way that it came apart.

Clean the transmission case with a steam cleaner, degreaser, or cleaning solvent. As you begin to disassemble the unit, pay close attention to the condition of its parts. Using a dial indicator, measure and record the end play of the input and main shafts. This information will be needed during assembly of the unit to select the appropriate selective shims and washers. Clean the transmission parts with solvent before assembly.

Task B.6   **Inspect, repair, and/or replace transmission shift cover and internal shift forks, bushings, bearings, levers, shafts, sleeves, detent mechanisms, interlocks, and springs.**

Inspect shift rails to be sure that they are not bent. A bent shift rail will not cause the transmission to jump out of gear or result in a gear clash or a gear noise.

Hard shifting may cause a bent shift rail. The shift rail is linked to the shifter handle. If the rail is bent, it may interfere with other parts in the transmission, causing a hard-shift feeling when the shift lever is moved from certain gears.

A linkage that is bent and a linkage that is out of adjustment are both common causes of a transmission that does not shift properly. Failure to go into gear is more commonly caused by a broken shift fork than by a damaged gear.

The shift linkages internal to the transmission are located at the top or side of the housing. Mounted inside the transmission is the control end of the shifter and the shift controls. The shift controls consist of the shift rail and the shift fork. As the shift fork moves toward the preferred gear, it moves the synchronizer sleeve to lock the speed gear to the shaft.

Task B.7   **Inspect and replace input (clutch) shaft, bearings, and retainers.**

A worn pilot bearing contact area on the input shaft or a worn pilot bearing could result in noise with the clutch pedal depressed. The main shaft is not turning with the engine idling, the clutch released, and the transmission in neutral. Because the input shaft is not turning with the clutch released, a rough input shaft roller bearing or needle bearings would not result in a growling noise under this condition.

An input shaft only uses one type of bearing: a ball bearing located toward the front half of the shaft (normally a pressed fit to the shaft). The bearing is lubricated by the fluid in the transmission. A needle bearing would not support the load to which an input shaft is subjected.

Task B.8   **Inspect and replace main shaft, gears, thrust washers, bearings, and retainers/snap rings; measure gear clearance/end play.**

A worn first-speed gear blocking ring or synchronizer sleeve may cause hard shifting, but would not result in noise while driving in first gear. A worn, rough main shaft bearing would cause a growling noise in all gears. Chipped and worn first-speed gear teeth would cause a growling noise in first gear.

The main shaft is not drilled with oil journals. Inspect the bearing surfaces of the Main shaft; it should be smooth and show no signs of overheating. Also inspect the gear journal areas on the shaft for roughness, scoring, and other defects. Check the shaft splines for wear, burrs, and other conditions that would interfere with the slip yoke's ability to slide smoothly on the splines.

Task B.9   **Inspect and replace synchronizer hub, sleeve, keys (inserts), springs, and blocking (synchronizing) rings/mechanisms; measure blocking ring clearance.**

The blocking ring dog teeth tips should be pointed with smooth surfaces. Clearance between the blocking ring and the matching gear dog teeth is important for proper shifting. The synchronizer sleeve must slide freely on the synchronizer hub. The threads on the blocking ring in the cone area must be sharp to get a good bite on the gear to stop it from spinning and to make a synchronized non-clashing shift.

If the clearance between the blocking ring and the fourth-speed gear dog teeth is less than specified, the blocking ring is worn, which results in hard shifting. This problem would not result in noise while driving in fourth gear.

Task B.10   **Inspect and replace countershaft, counter (cluster) gear, bearings, thrust washers, and retainers/snap rings.**

Because the counter (cluster) gear is turning with the clutch pedal released in neutral and in gear, damaged counter (cluster) gear bearings may cause a growling noise with the engine idling with the transmission in neutral and the clutch pedal released. It will also cause a growling noise while driving in any gear.

All counter (cluster) gears should show wear patterns in the center of their teeth. These wear patterns should appear as a polished finish with little wear on the gear face. Check the gears' teeth carefully for chips, pitting, cracks, or breakage. Also, inspect the bearing surfaces to make sure they are smooth. Any damage to the assembly requires replacement.

Task B.11   Inspect and replace reverse idler gear, shaft, bearings, thrust washers, and retainers/snap rings.

Because the reverse idler gear is only in mesh with reverse gear, this gear rotates in reverse gear only.

Inspect the reverse idler gear for pitted, cracked, nicked, or broken teeth. Check its center bore for a smooth surface. Carefully inspect the needle bearings on which the idler gear rides for wear, burrs, and other defects. Also, inspect the reverse idler gear's shaft surface for scoring, wear, and other imperfections. Replace any part that is damaged or excessively worn.

A worn extension housing bushing may cause premature extension housing seal wear and fluid leaks. Excessive main shaft end play has no effect on speedometer operation.

Task B.12   Measure and adjust shaft/gear, and synchronizer end play.

In fourth gear, the 1-2 synchronizer is moved ahead so the synchronizer hub is meshed with the dog teeth on the fourth-speed gear on the input shaft. Excessive input shaft end play would cause the transmission to jump out of fourth gear.

While disassembling a transmission, the technician should be taking end play readings. These readings will be recorded. When the transmission is reassembled, select-fit thrust washers and shims will be used to set all parts to specifications. A main shaft should not be replaced unless, during inspection, excessive wear or damage is found. Bearings and snap rings can be reused if nothing is found during cleaning and inspection.

Task B.13   Measure and adjust bearing preload or end play.

Bearing preload is adjustable. It is normally adjusted by placing force on the bearing using a select-fit shim behind the bearing or by an adjusting nut that applies pressure to the bearing. Bearing preload is used to take play out of the bearing so the bearing takes the load correctly. Packing a bearing is not part of setting bearing preload.

A worn speedometer drive and driven gears may cause erratic speedometer operation, but this problem would not result in premature extension housing bushing failure. A plugged transmission vent may cause excessive transmission pressure and fluid leaks, but this problem would not affect extension housing bushing wear. Excessive main shaft end play causes lateral shaft movement, but this does not cause excessive extension housing bushing wear. Metal burrs between the extension housing and the transmission case cause misalignment of the extension housing, which forces the extension housing bushing against the driveshaft yoke in one location, resulting in bushing wear.

Task B.14   Inspect, repair, and replace extension housing and transmission case mating surfaces, bores, bushings, and vents.

Check the extension housing for cracks and repair or replace it as needed. Check the mating surfaces of the housing for burrs or gouges and file the surface flat. Inspect the speedometer cable or sensor for any leakage from the seal; replace the seal if fluid is leaking. Install a new gasket to the extension housing during installation. Check all threaded holes and repair any damaged bores with a thread repair kit. Check the bushing in the rear of the extension housing for excessive wear or damage. Always replace the rear extension housing seal.

Task B.15   Inspect and replace speedometer drive gear, driven gear, and retainers.

The speedometer driving gear is the gear located on the output shaft of the transmission and can be accessed by removing the rear extension housing. The speedometer driven gear is located on the vehicle speed sensor and is located in the rear extension housing. There are no speedometer gears located on the main shaft or the axle shaft.

Task B.16   Inspect, test, and replace transmission sensors and switches.

Manual transmissions may have various sensors, switches, and solenoids. These components may include a vehicle speed sensor (VSS), a backup light switch, and a computer-aided gear select solenoid. The VSS sends a voltage signal to the PCM in relation to vehicle speed. The PCM uses this signal to operate output such as the cruise control. Some vehicles multiplex sensors and obtain the VSS signal from ABS wheel speed sensors. The ABS/VSS sensor is usually a permanent-magnet signal generator.

However, some units utilize a Hall-effect device to generate the voltage signal. While some of these types of sensors may be checked using an ohmmeter for initial resistance, the only accurate method of testing the sensors is by using a DMM on the AC voltage setting or by using an oscilloscope.

The backup light switch is operated by linkage inside the transmissions to operate the backup lights when the transmission is placed in reverse. The switch is normally open and has power going to it when the vehicle ignition is on. The backup lamp lights when the vehicle is shifted into reverse. The switch will close when the vehicle is shifted in reverse and provide a path to ground for the backup lamps to operate. The computer-aided gear select solenoid ensures good fuel economy and compliance with federal fuel economy standards by inhibiting second and third gears when shifting out of first gear under certain conditions.

Task B.17   Inspect lubrication systems.

Transmission fluid level must be maintained at the level of the check plug in the transmission case or at a level marked on a transmission dipstick. Many late-model manual transmissions and transaxles use automatic transmission fluid (ATF) as a lubricant for reduced friction and improved vehicle fuel economy. Some manual transmissions use hypoid gear oil as a lubricant, and a few use motor oil.

The hypoid ring-and-pinion gearsets in rear-wheel drive axles require hypoid gear oil, usually GL4 or GL5. Limited-slip differentials require additional fluid additives. The viscosity of hypoid gear oil is higher (thicker) than that of motor oil or ATF. It may be single-viscosity, such as SAE 90, or multiple-viscosity, such as 85W-90. Many final-drive gearsets in front-wheel drive (FWD) transaxles are not hypoid gears and use ATF or motor oil as a lubricant. Some FWD final drives are hypoid gearsets, however, and require GL4 or GL5 gear oil. Always follow the manufacturer's specifications for fluid type, viscosity, and replacement intervals.

Task B.18   Check fluid level and refill with proper fluid.

It is important for technicians to include an inspection of drivetrain fluid level and to be sure that manufacturers' recommended types of fluids are used. As a technician, you should check the manufacturer's TSBs for changes in fluids or additives that have been recommended since the production date. When fluid level is incorrect, it should be adjusted with the correct fluid type to proper levels.

C. Transaxle Diagnosis and Repair (8 Questions)

Task C.1   Diagnose transaxle noise, hard shifting, gear clash, jumping out of gear, fluid condition, and fluid leakage problems; determine needed repairs.

A worn fourth-speed synchronizer will only affect shifting in fourth gear. Excessive main shaft end play may result in gear clash in all gears. A worn 3-4 shift fork may cause shifting problems in third and fourth gear, but this problem will not result in gear clash in all gears. A clutch disc sticking on the input shaft will cause the clutch not to release properly, resulting in gear clash in all gears.

A worn blocker ring, damaged speed gear, or a worn bushing could cause gear clash in a particular gear. Gear clash in all gears could be caused by stretched shifter cables.

Task C.2   Inspect, adjust, lubricate, and replace transaxle external shift assemblies, linkages, brackets, bushings/grommets, cables, pivots, and levers.

A misadjusted shift linkage may cause many problems. If the linkage is misadjusted, the transmission may not be able to be shifted all the way into gear. This will cause further damage to the transmission. An improper shift linkage adjustment also may cause hard shifting or sticking in gear.

Most transmissions and transaxles are adjusted with the unit in neutral. A 1/4-inch (6.35 mm) bar or drill bit is installed in the lever to hold the transaxle in neutral while the cables or linkage are adjusted. Transmissions and transaxles with internal linkage have no adjustment. Shift cables should not be modified in any way, only replaced or adjusted.

Task C.3   **Inspect and replace transaxle gaskets, sealants, seals, and fasteners; inspect sealing surfaces.**

A worn outer drive axle joint may cause a clicking noise while cornering at low speed, but this defect would not cause repeated drive axle seal failure. A plugged transaxle vent may cause excessive transaxle pressure and repeated drive axle seal failure.

Transaxle cases have machined mating surfaces that have a very smooth flat finish on them. Not all require a gasket, but they do require some sort of sealant that should be equivalent to the manufacturer's specifications. If a transaxle is assembled without following manufacturer's sealing instructions, leakage from the case will result.

Inspect the transaxle mating surfaces for warpage with a straightedge before assembly to ensure a proper fit.

Task C.4   **Remove and replace transaxle; inspect, replace, and align transaxle mounts, and subframe/cradle assembly.**

A misaligned engine and transaxle cradle may cause drive axle vibrations. Because the lower control arms are connected to this cradle, misalignment of the cradle may cause improper front suspension angles.

When removing a transaxle from a vehicle, it is necessary to install an engine support bracket. This will hold the weight of the engine while the transaxle is being removed. It is installed for the safety of the technician, as well as to avoid damaging the vehicle. You do not have to drain engine oil when removing a transaxle. You should disconnect the negative battery cable. Not all vehicles require you to remove the engine along with the transaxle.

Task C.5   **Disassemble and clean transaxle components; reassemble transaxle.**

While assembling a manual transaxle, it is important to apply gear lube to all of the transaxle parts. Before checking the specifications of the shafts in the transaxle, rotate the shafts to work the gear lube into the bearings. If the gear lube is not worked into the bearings a false measurement may be made.

Before disassembling a transaxle, observe the effort it takes to rotate the input shaft through all forward gears and reverse. Extreme effort in any or all gears may indicate an end play problem or a bent shaft.

Task C.6   **Inspect, repair, and/or replace transaxle shift cover and internal shift forks, levers, bushings, shafts, sleeves, detent mechanisms, interlocks, and springs.**

Worn dog teeth on the fourth-speed gear would not cause the transaxle to jump out of third gear. A weak detent spring on the 3-4 shift rail may cause the transaxle to jump out of third gear. By not having enough spring pressure, the weak spring could cause this to happen.

The shift forks are used to shift gears, but they are not connected to the forward gears or reverse gear. Also, the shift forks are not connected to the blocker rings or the countershaft. A blocker ring is used to stop a gear from spinning while the vehicle is in motion for gear synchronization. A countershaft is used to change the rotation of the gears on the main shaft and for different gear ratios. The shift forks are connected to the synchronizer assembly. They move the synchronizer sleeve forward or backward to engage the transaxle in a gear.

Task C.7   **Inspect and replace input shaft, gears, bearings, and retainers/snap rings.**

Worn dog teeth on the third-speed gear or blocking ring may cause hard shifting or jumping out of third gear, but this problem would not cause a growling noise while driving in third gear. Worn threads in a third-speed blocking ring may cause hard shifting, but this wear would not cause a growling noise in third gear. Worn, chipped teeth on the third-speed gear could result in a growling noise while driving in third gear.

Task C.8   Inspect and replace output shaft, gears, thrust washers, bearings, and retainers/snap rings.

Worn dog teeth on the second-speed gear and blocking ring would not result in a growling noise while driving or accelerating in second gear, but this problem could not cause hard shifting in second gear. Worn dog teeth on the second-speed gear and blocking ring may cause the transaxle to jump out of second gear.

Inspect all small parts in the transmission for wear. A service manual may list specifications for the thickness of parts, such as thrust washers. If specifications are not available, inspect each part for signs of wear or breakage. Normally all the snap rings, roller bearings, washers, and spacers are replaced during a transaxle overhaul. Most manufacturers sell a small parts kit that includes all of these parts.

Task C.9   Inspect and replace synchronizer hub, sleeve, keys (inserts), springs, and blocking (synchronizing) rings; measure blocking ring clearance.

Before disassembly, always mark the synchronizer sleeve and hub so that these components can be reassembled in their original locations. Synchronizer hubs are not reversible on the shaft. Synchronizer sleeves are not reversible on their hubs.

A worn blocker ring will cause the transmission/transaxle to have gear clash. If the blocker ring is worn in the cone area (meaning that all of the sharp ridges are dull or gone), the blocker ring will not work properly. A blocker ring should stop a gear from spinning through the sharp ridges in the cone area before the synchronizer sleeve engages the gear.

If the drive gear is not positioned correctly on the output shaft, it will not operate. Stripped driven gear teeth is a common cause of an inoperative speedometer. Driven gear slippage on the bottom of the VSS is a common cause of a bad driven gear. Although not very common, the drive gears may be stripped, causing an inoperative speedometer. A mispositioned drive gear is a very unlikely cause of an inoperative speedometer.

Task C.10   Inspect and replace reverse idler gear, shaft, bearings, thrust washers, and retainers/snap rings.

Inspect the reverse idler gear teeth for chips, pits, and cracks. Although worn reverse idler teeth may cause a growling noise while driving in reverse, this problem would not cause a failure to shift into reverse. A broken reverse shifter fork may not allow the transaxle to shift into reverse without causing noise.

The needle bearings on a reverse idler gear should be smooth and shiny. Carefully inspect the needle bearings for wear, burrs, and other problems. Replace worn or damaged needle bearings or damage to other components may occur.

Task C.11   Inspect, repair, and/or replace transaxle case mating surfaces, bores, dowels, bushings, bearings, and vents.

Transaxle case replacement is often required if the case is cracked. Some vehicle manufacturers recommend that the case be repaired, depending on how extensive the damage. The transaxle case may be repaired with an epoxy-based sealer for some transaxle cracks. Loctite® is not recommended for repairing any transaxle case.

If a threaded area in an aluminum housing is damaged, use service kits to insert new threads in the bore. Some threads should never be repaired; check the service manual to identify which ones can be repaired.

Task C.12   Inspect and replace speedometer drive gear, driven gear, and retainers.

A speedometer gear is normally mounted on the output shaft. The output shaft spins at driveline speed. An output shaft will never have a drive gear machined into the output shaft. A drive gear is made out of a plastic nylon-type gear, and the teeth on the drive and driven gear have a helical-type cut on them. The helical-type cut and the plastic-type gear are used for quiet operation.

Whenever a speedometer drive gear is replaced, the driven gear also should be replaced. If a speedometer cable assembly core is damaged, it may be replaced with a new core. The new core must be cut to the same size as the one being replaced and be properly lubricated before installation into the cable housing.

Task C.13   Inspect, test, and replace transaxle sensors and switches.

Most VSS are magnetic pickup coil signal generators. Some from the early 1980s, however, are rotary magnetic switches or optical sensors. The signal from a magnetic pickup VSS is an analog sine wave that varies in frequency and amplitude with vehicle speed. The signal from a magnetic switch or an optical VSS is a digital square wave that varies in frequency only. The signal integrity and waveform of any kind of VSS is best tested with an oscilloscope.

Task C.14   Diagnose differential assembly noise and vibration problems; determine needed repairs.

Damaged ring gear teeth would cause a clicking noise while the vehicle is in motion. This problem would not cause differential chatter. Improper preload on differential components, such as side bearings, may cause differential chatter.

If there were a constant whining noise coming from the differential, the noise could not be coming from the side gears or the spider gears. These gears are only used when the vehicle is turning; so if they were damaged, the noise would only be heard on turns. The wrong differential lube could cause damage to the differential parts, but will not cause a whining noise. If the preload and backlash are not set properly, the gear mesh could be too tight and cause a whining noise.

Task C.15   Remove and replace differential final drive assembly.

In most transaxles, the differential is serviced or removed by disassembling the halves of the transaxle case. The differential bearings fit into the case and usually are not adjustable for bearing preload. If the side bearings allow for preload, there are shims that fit between the transaxle housing and the bearing cup or assembly. The differential can usually be removed without complete disassembly of the transaxle gear train.

Task C.16   Inspect, measure, adjust, and replace differential pinion gears (spiders), shaft, side gears, thrust washers, and case.

The side gear end play must be measured individually on each side gear with the thrust washers removed. Side gears with the specified thrust washer have slight end play, but no preload.

The spider gears ride on the pinion shaft. The bore in the gears should be smooth, shiny, and have no signs of pits or scuffing. The pinion shaft should also be free of pitting and scoring. There is no needle bearing in any of the spider gears.

Task C.17   Inspect and replace differential side bearings; inspect case.

A side bearing preload adjustment shim is positioned behind one of the side bearing cups. When the differential side bearings are being replaced, they do not need to be packed with grease. Lubricate the bearings with the differential lube. The differential case does not need to be replaced when the bearings are replaced. The only time a case is replaced is when it shows signs of damage. When new bearings are installed, they should be installed using a hydraulic press, and the bearing races should be replaced also. The new bearing will not wear in properly if the old race is used.

Most late-model vehicles use an ATF or a motor oil of 5W30 or 10W30 in manual transaxles. Late-model vehicles require a thinner, more free-flowing fluid because tolerances are a lot tighter than they used to be. A thicker fluid may not provide enough protection with such close or tight tolerances. Also, late-model vehicles need to meet certain gas mileage levels without giving up performance. Thinner transmission fluid will allow for less load on the engine, helping gas mileage and performance.

Task C.18   Measure shaft preload/end play (shim/spacer selection procedure).

While measuring the differential side play to determine the required side bearing shim thickness, a new bearing cup is installed in the transaxle case without the shim. The proper shim thickness is equal to the differential end play plus a specified thickness for bearing preload. While measuring the differential end play, tighten the transaxle case bolts to the specified torque. Apply a medium load to the differential in the upward direction while measuring the differential end play.

When the differential turning torque is less than specified, increase the shim thickness behind the side bearing cup in the bell housing side of the transaxle.

Task C.19   Inspect lubrication systems.

Wrong transaxle lubricant may cause burned output shaft bearings. A broken oil feeder behind the front output shaft bearing results in improper output shaft bearing lubrication and burned bearings. If a bearing is not lubricated properly, heat will build and destroy the bearing and possibly other parts of the transaxle.

During transmission or transaxle overhaul, drain and inspect the fluid. Gold-color particles in the fluid are from the wearing of the brass blocking rings on the synchronizers. Metal shavings in the fluid are produced from the wearing of the gears. An excessive amount of shavings in the fluid indicates severe gear and synchronizer wear.

On a cold start, if the fluid is too thick, the vehicle may be hard to shift. This will also cause the transaxle to get poor lubrication on a cold start.

Task C.20   Check fluid level and refill with proper fluid.

It is important for technicians to include an inspection of drivetrain fluid level and to be sure that manufacturer's recommended types of fluids are used. As a technician, you should check manufacturer's TSBs for changes in fluids or additives that have been recommended since the production date. When fluid level is incorrect, adjust it with the correct fluid type to proper levels.

Task C.21   Measure and adjust differential bearing preload/end play.

Improper adjustment of differential tapered bearing preload can cause differential noises and premature component failures. It should be measured and adjusted as per manufacturer's recommended procedure and specification.

D. Drive Shaft/Half-Shaft and Universal Joint/ Constant Velocity (CV) Joint Diagnosis and Repair (Front and Rear Wheel Drive) (5 Questions)

Task D.1   Diagnose shaft and universal/CV joint noise and vibration problems; determine needed repairs.

A worn universal joint (U-joint) may cause a squeaking noise that increases in relation to vehicle speed. If the centering ball and socket is worn in a double Cardan U-joint, a heavy vibration may occur during acceleration.

A torsional damper will not cause a clicking noise if it is worn out or bad. If the torsional damper was bad, a shudder would be felt in the vehicle. A constant-velocity (CV) inner joint is not affected when the vehicle is turning. It is only affected when the vehicle suspension is jounced or rebounded while driving. Axle shafts are serviceable; therefore, the whole axle shaft does not need to be replaced. An outer CV joint will make a clicking noise when the vehicle is turning. This means the joint ball bearings are bad and the grease is contaminated. The CV joint should be replaced.

Task D.2   Inspect, service, and replace shafts, yokes, boots, and universal/CV joints; verify proper phasing.

A worn front wheel bearing usually results in a growling noise while cornering or driving straight ahead. A clunking noise while decelerating may be caused by a worn inner drive axle joint.

A universal joint has prelube on the inside. However, this prelube is not sufficient to lubricate the joint when it has been installed in a vehicle. You should grease a new universal joint when it is installed. When greasing a universal joint, you should not pump so much grease into the joint that it squirts out of the caps. When this happens, it damages the seals around the caps and shortens the life of the joint.

Task D.3   Inspect, service, and replace shaft center support bearings.

A worn driveshaft center bearing or an outer rear axle bearing causes a growling noise that is not influenced by acceleration and deceleration.

A center support bearing is usually maintenance-free and seated. A center support bearing is not part of the driveshaft, but is part of the driveline. A center support bearing is found mainly on trucks and vans. It is used to shorten the length of long drive shafts and to decrease pinion angles.

Task D.4   Check and correct drive/propeller shaft balance.

Before checking driveshaft balance, always inspect the shaft for damage. A missing balance weight, accumulation of dirt, or excessive undercoating will affect the balance.

To check driveshaft balance, chalk mark the driveshaft at four locations 90 degrees apart slightly in front of the driveshaft balance weight. Hold a strobe light against the rear axle housing just behind the pinion yoke. Run the vehicle in gear until the driveshaft vibration is the most severe. Point the strobe light at the chalk marks on the driveshaft and note the position of one reference mark. The number that appeared on the strobe light should have two screw-type clamps installed on the shaft near the rear with the heads of the clamp opposite the number that appeared. The vehicle's suspension should have the weight of the vehicle on it when this procedure is performed so that the suspension is at its normal ride height and there are no abnormal pinion angles.

Task D.5   Measure driveshaft runout.

While measuring driveshaft runout, position the dial indicator near the center of the driveshaft. Replace the driveshaft if the runout is excessive. A driveshaft that is bent or damaged in any way should be replaced; repairs to a damaged driveshaft should not be attempted.

To check driveshaft runout, raise the vehicle and install a dial indicator with a magnetic base under the vehicle near the center of the driveshaft. The surface of the shaft should be wiped off or cleaned in case it is rusted or has dirt buildup that may affect the reading on the dial indicator. Rolling a driveshaft on a flat surface is not an accurate way of checking the driveshaft runout.

Task D.6   Measure and adjust driveshaft working angles.

Driveshaft working angle is also known as pinion angle. The engine and transmission are installed in the chassis at a preset angle usually pointing down at the tailshaft in rear-wheel drive applications. For our purposes, assume the tailshaft is pointing down three degrees. To avoid a humming or droning vibration in the driveline, the pinion must be pointing up three degrees. This creates a situation where the planes on which they operate in are parallel. It also provides the smoothest U-joint operation. (Note that most manufacturers only give specs for pick-up trucks for this measurement. Unless the manufacturer specifies differently, the static setting of each should be the same.) Causes of driveshaft working angle problems are: sagging (damaged) rear springs, sagging engine or transmission mounts, or major changes in ride height up or down.

Task D.7   Inspect, service, and replace front wheel bearings, seals, and hubs.

Wheel bearings should be cleaned, inspected for wear or damage, and packed with the appropriate wheel bearing grease using either a packing tool or the hand-pack method.

If only one bearing is bad, replace only that bearing. When a wheel bearing is replaced, replace the bearing race also. The new bearing may fail if the race is not replaced.

E. Rear-Wheel Drive Axle Diagnosis and Repair (7 Questions)

Task E.1   Ring and Pinion Gears (3 Questions)

Task E.1.1   Diagnose noise, vibration, and fluid leakage problems; determine needed repairs.

Noises in differentials can be from the axle bearings, ring and pinion, any of the differential bearings, or differential side gears. Axle bearings usually can be isolated by the change in noise as the vehicle experiences different side loads or by raising the vehicle and running it in gear. Most axle bearing noises will subside dramatically when weight is taken off the wheels. Ring and pinion noises are associated with a whine or growl that changes in pitch as vehicle speed or engine load changes. To diagnose differential carrier and pinion bearing noises, a stethoscope is often employed to pinpoint the location of the noise. Because the side gears are only turning while cornering, they do not cause a whining noise while driving straight ahead. If the differential fluid is too full, excessive pressure may build up and cause the differential fluid to leak past a seal. If the vent becomes plugged, it will cause excessive pressure in the differential housing and a leak will occur. Axle shaft bearings that are worn will cause the axle shaft to apply load on the axle shaft seal, and the seal will fail. Pinion seals and carrier covers are some other common and easy to spot leak points.

Task E.1.2   Inspect and replace companion flange and pinion seal; measure companion flange runout.

To check pinion flange runout, remove the driveshaft and mount a dial indicator against the face of the flange. Rotate the flange and note the readings on the dial indicator; these are the runout readings. If the flange is removed, you cannot measure pinion flange runout. A loose pinion nut allows pinion shaft end play, resulting in a clunking noise on acceleration and deceleration. Insufficient pinion nut torque will affect the pinion bearing. Preload will not cause a growling noise.

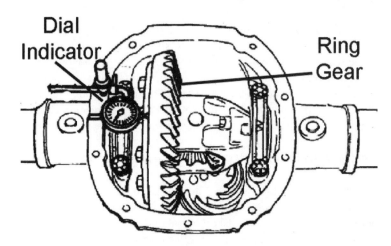

Task E.1.3   Measure ring gear runout; determine needed repairs.

Ring gear runout is best measured before differential disassembly. In most Applications, a dial indicator is mounted to the carrier or axle housing and the measuring tip is set on the back of the ring gear (the side without gear teeth). Runout is measured by turning the ring gear and watching for variations in the dial indicator. Use manufacturer specs to determine if the gear has excessive runout. Possible causes for ring gear distortion are uneven torque of ring gear bolts, overheating of gear assembly (usually evident long before you take this measurement), the gear was dropped during assembly, manufacturing defects, and debris between the carrier and the gear. Torque and debris issues can usually be corrected without replacing the gear, but all other issues require ring and pinion set replacement.

Task E.1.4   Inspect and replace ring and pinion gear set, collapsible spacers, sleeves (shims), and bearings.

This task covers an enormous area. Much of the related information in this task is in other areas of this section of the task list. If you are not familiar with the process of rebuilding a differential, it would be wise to read through any manufacturer's procedure as that task exceeds the scope of this book. Key issues that you should know are: the ring and pinion set must be replaced as a set; collapsible spacers must be replaced if they are over tightened or when the pinion bearings need to be replaced; and bearings should be replaced as assemblies. When determining the condition of a ring and pinion, you are looking for excessive wear on the gears, pitting or grooves worn in the gear faces, or evidence that the gears have overheated which will cause them to turn blue or black.

Task E.1.5   Measure and adjust drive pinion depth.

The primary purpose of adjusting the pinion gear depth is to set the pinion and ring gear mesh. Pinion depth setting is the distance between a point—usually the gear end—of the pinion gear and the centerline of the axles or differential case bearing bores. The pinion gear depth is normally adjusted by installing shims onto the pinion mounting.

These measurements must be taken with the pinion bearings pre-loaded.

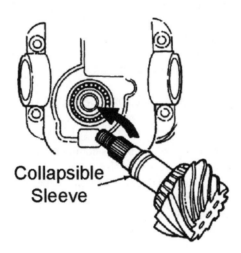

Collapsible
Sleeve

Task E.1.6   Measure and adjust drive pinion bearing preload (collapsible spacer or shim type).

When replacing ring and pinion sets, the first item to be installed and adjusted is the pinion gear. It has a bearing on the gear side and another on the companion flange side. The bearings must be preloaded to keep them from deflecting under the sideways loads the driveshaft and ring gear put on it. Preload is set either using shims between the bearings or a collapsible collar. Preload is measured in inch-pounds of resistance when turning the pinion nut. The pinion nut must never be loosened to obtain the specified turning torque. Lubricate the pinion bearings when the turning torque is measured.

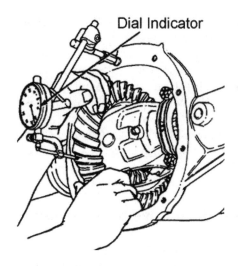

Dial Indicator

Task E.1.7   Measure and adjust differential (side) bearing preload and ring and pinion backlash (threaded adjuster or shim type).

Ring gear runout and case side play should be measured before removing the ring gear and case assembly. Mark the side bearing caps in relation to the case before removal. Lubricate the side bearings before installation.

The side bearings must be in good condition before measuring case runout. The ring gear runout should be measured before the case runout.

Task E.1.8   Perform ring and pinion tooth contact pattern checks; determine needed adjustments.

Prussian blue (a marking compound) is commonly used to check the tooth contact pattern on gear setup. A used gearset will have a shiny pattern on the gear teeth that can be visually inspected. Look for the pattern to be centered with either new or used gears after assembly. Carrier-bearing adjustment may be necessary to correct the depth and backlash of the gears. Do not concern yourself with the exact appearance of the pattern as it varies by gear vendor. Take some time to look at the location of the pattern on the ring and pinion faces for your favorite manufacturer.

Task E.2   Differential Case/Carrier Assembly (2 Questions)

Task E.2.1   Diagnose differential assembly noise and vibration problems; determine needed repairs.

Differentials perform their function only when a vehicle is turning a corner. Noises with any differential will usually be present only when turning unless there is a significant difference in tire dimensions from one side to the other. Grinding or whining noises are associated with differential side gears. Popping or crunching sounds usually are associated with differential cross-shaft failure and gear failure.

Task E.2.2   Remove and replace differential assembly.

On some types of axles, the pinion gear and differential assembly all come out of the axle housing as one assembly. Always mark the bearing caps when removed to ensure they go back together correctly. The axle shafts must be removed before the differential assembly will come out. The shim packs and bearing races should be kept in order.

To get a rotating torque measurement on the pinion gear, remove the differential case. If the load of the case, ring gear, and the axle shafts are included, you measure the rotating torque of the axle assembly. Use an inch-pound torque wrench with a needle-type indicator to get an accurate reading.

Task E.2.3   **Inspect, measure, adjust, and replace differential pinion gears (spiders), shaft, side gears, thrust washers, and case/carrier.**

Use feeler gauges to check the clearance between differential side gears and thrust washers. Remove the differential side gears and pinions. Inspect all gears for abnormal wear, heat damage, chipped teeth, brinelling, and other wear. Inspect the differential pinion shaft for wear, cracks, lack of lubrication, and heat damage. When assembling differential gears and side gears, place the side gears and washers into the case. Then walk the pinions around the side gears until they align with the shaft hole; insert the shaft, spacer block, and lockpin. After installation, rotate the gears a few times and recheck thrust washer clearance with feeler gauges.

Task E.2.4   **Inspect and replace differential side bearings; inspect case/carrier.**

Inspect the side bearings for wear, particularly signs of the inner races turning on the bearings. Replace worn parts. Inspect thrust washer surfaces for nicks and abnormal wear. If new side bearings are required, lubricate the surfaces that contact the case.

Install shims if required and drive or press the bearings onto the case. Apply force on the inner cone, not on the rollers. If new bearings are installed, install new outer races (cups) in the axle housing or differential carrier.

Task E.2.5   **Measure differential case/carrier runout; determine needed repairs.**

To check differential case runout, place the case in a set of V-blocks. Then place a dial indicator against the ring gear mounting flange and rotate the case to measure axial runout. Move the dial indicator pointer to the ring gear hub area of the case and rotate the case again to measure radial runout. If runout is out of limits in either direction, replace the case.

Task E.2.6   **Inspect axle housing and vent.**

During differential inspection, be sure to inspect the axle housing for signs of damage and fluid leakage. Be sure to inspect vent for blockage as this can cause fluid leakage and result in low fluid levels that can damage internal parts.

Task E.3   **Limited Slip/Locking Differential (1 Question)**

Task E.3.1   **Diagnose limited slip differential noise, slippage, and chatter problems; determine needed repairs.**

Differentials only perform their function when a vehicle is turning a corner. Noises with any differential will usually be present only when turning unless there is a significant difference in tire dimensions from one side to the other. Limited-slip differentials are designed to lock the two wheels together via the differential when accelerating in a straight line. Open or non-limited-slip differentials only drive one wheel during acceleration. Locking of the differential is accomplished through the use of clutches or spring-loaded plates that will release and slip during cornering. Noises, slippage, and chatter usually occur due to improper fluid in the axle or wear of the limited-slip components. Special additives are used in most limited-slip differentials to help provide smooth release and engagement of the differential components.

Task E.3.2   **Inspect, drain, and refill with correct lubricant.**

When filling a transmission or differential case, fluid will come out of the fill hole as it reaches the full level. After draining differential fluid, inspect it for excessive metal particles. Silver- or steel-colored particles are signs of gear or bearing wear. Copper- or bronze-colored particles are signs of limited-slip clutch disc wear.

Task E.3.3   **Inspect, adjust, and replace clutch (cone/plate) pack or locking assembly.**

The friction plates have a minimum thickness specification and should be measured with a micrometer. The friction plate has to be removed from the clutch pack for this measurement. There is no way to measure a friction plate with a feeler gauge.

Task E.4   Axle Shafts (1 Question)

Task E.4.1   Diagnose rear axle shaft noise, vibration, and fluid leakage problems; determine needed repairs.

Review E.1.1 for bearing noise information. Many axle shafts ride directly on the axle bearing and seem to be more susceptible to wear which results in noise and fluid leaks.

Task E.4.2   Inspect and replace rear axle shaft wheel studs.

When replacing a wheel stud, use a hammer to carefully tap the old stud out. When installing a wheel stud, use an installation tool to avoid damage to the new stud. A torch should never be used to burn out a wheel stud because damage to the axle shaft and axle seal may occur.

Task E.4.3   Remove, inspect, and/or replace rear axle shafts, splines, seals, bearings, and retainers.

A worn axle shaft bearing can cause axle shaft seal failure. When the bearings wear out, axle side movement applies a greater load on the seal lip. Scored axle shafts in the seal area will damage the seal lip and cause the seal to fail. The axle shaft seal comes with a sprayed-on sealant and does not require any other sealant. Lubricate the sealing lip of the axle shaft seal with a light coating of gear lube to prolong the seal life.

If the axle shaft is damaged near the seal area, discard the shaft and replace it with a new one.

Task E.4.4   Measure rear axle flange runout and shaft end play; determine needed repairs.

To measure the axle shaft end play, you will need to remove the wheel and tire assembly and the brake drum. A dial indicator is mounted or clamped to the axle housing or suspension. Push the axle shaft in to the housing all the way until it stops. Rest the dial indicator head on the face of the axle shaft flange. With the dial indicator set to zero, pull out on the axle shaft. The resulting dial indicator reading is the axle shaft end play. The vehicle differential cover does not need to be removed.

Excessive runout could be caused by a bent axle shaft. A worn C-lock will cause excessive end play. A worn bearing will cause fluid leakage. A bent housing is considered major damage and is noticeable.

F.   Four-Wheel Drive/All-Wheel Drive Component Diagnosis and Repair (7 Questions)

Task F.1   Diagnose drive assembly noise, vibration, shifting, and steering problems; determine needed repairs.

Worn U-joints may cause a squeaking or clunking noise, and a vibration while driving straight ahead. Worn outer front drive axle joints on a 4WD vehicle may cause a vibration while cornering.

When a vacuum-operated 4WD does not shift into 4WD, the engine vacuum may be low or the vacuum motor at the front differential may be damaged. Another cause could be bad or disconnected vacuum lines.

Task F.2   Inspect, adjust, and repair transfer case manual shifting mechanisms, bushings, mounts, levers, and brackets.

Specification measurements should be taken and recorded to aid in the installation of parts that have tolerances. Clean and lubricate all parts before the assembly of the part. Inspect all components for wear or damage.

The annulus gear is locked to the case so it cannot rotate. In 4WD low, the transfer case input shaft is driving the sun gear, which, in turn, is driving the planetary carrier.

Task F.3 Remove and replace transfer case.

Transmission and transfer cases are removed as an assembly. Most newer 4WD vehicles have the transfer case bolted to the rear of the transmission. Not all transfer cases are made of cast iron; most new models have a lightweight aluminum case.

Task F.4 Disassemble transfer case; clean and inspect internal transfer case components; determine needed repairs.

A plugged transfer case vent may cause seal leakage. A remote transfer case vent helps keep water out of the transfer case.

Task F.5 Reassemble transfer case; refill with proper fluid.

If the drive chain in the transfer case is stretched, and the drive sprocket teeth are worn, the chain could slip on the teeth and cause a loud clicking noise under acceleration.

If the four-wheel drive (4WD) on the vehicle is not used often, the shift linkage could become rusted, and it will not work easily or at all. The shift linkage should be inspected for bushings that may be worn or deteriorated and need replacement. If the shift linkage does not move the linkage its full range, the transfer case may not operate in 4WD. The transfer case would still shift into gear if the front driveshaft universal joints are bound. Low fluid may cause damage to internal components, but the transfer case will still engage. A manual shift transfer case has no electronic shift motor.

Task F.6 Check transfer case fluid, level, condition, and type.

Task F.7 Inspect, service, and replace front drive/propeller shaft and universal/CV joints.

Worn universal joints (U-joints), front axle drive joints, and incorrect driveshaft angles may cause a vibration that is more noticeable when changing throttle position. A snap ring holds the inner tripod joint on the axle shaft, and a special swaging tool may be necessary to tighten the outer boot clamp. A worn outer constant-velocity (CV) 32 Manual Drive Train and Axles (Test A3) **Overview of the Task List** joint may cause a clicking noise while cornering. All of the grease supplied with the joint should be used; this is the correct amount to be used.

Task F.8 Inspect, service, and replace front drive axle, universal/CV joints and drive/half shaft.

It is not necessary to remove the axle assembly to remove the drive axles. A slide hammer may be needed to remove the drive axles from the housing.

Task F.9 Inspect, service, and replace front wheel bearings, seals, and hubs.

Neither the automatic locking hubs nor the caps should be packed with grease. If they are packed with grease, they will not operate properly; the parts must move freely. If only one bearing is bad, replace only that bearing. When you replace a wheel bearing, also replace the bearing race. The new bearing may fail if the race is not replaced.

Task F.10 Check transfer case and front axle seals and all vents.

The only purpose of a remote vent is to keep moisture out of a differential assembly. This is needed on a 4WD vehicle in case the axle or transfer case is submerged in water.

Task F.11 Diagnose, test, adjust, and replace electrical/electronic components of four-wheel/all-wheel drive systems.

An electronic-shift transfer case has all of the shift linkage inside the case. If an electronic-shift transfer case has a problem with shifting, an electrical component would be a likely cause of failure.

Task F.12   **Test, diagnose, and replace axle actuation and engagement systems (including viscous, hydraulic, magnetic, and mechanical).**

Electronically controlled actuation and engagement systems require special diagnostic test equipment. As a technician, you should have an understanding of electrical and electronic operational theory. Be sure to refer to manufacturer's recommended testing and service procedures. Damage to electronic components and control devices can result from improper testing and installation procedures.

5 Sample Test for Practice

Sample Test

Please note the letter and number in parentheses following each question. They match the task in Section 4 that discusses the relevant subject matter. You may want to refer to the overview using the cross-referencing key to help with questions posing problems for you.

1. A manual-shift transfer case will not shift into four-wheel drive (4WD). The cause of this problem could be:
   A. the fluid is low in the transfer case.
   B. the front driveshaft universal joints are bound up.
   C. the shift linkage needs lubrication.
   D. the electronic shift motor is bad. (F.5)

2. All of the following statements regarding four-wheel drive (4WD) front-drive axles and joints are true **EXCEPT:**
   A. the inner tripod joint is held on the axle shaft with a snap ring.
   B. a special swaging tool may be required to tighten the outer boot clamps.
   C. the new joint is coated with the grease supplied, and the remaining grease is discarded.
   D. a worn CV joint may cause a clicking noise while cornering. (F.6)

3. In an electronically shifted transfer case, the shift motor is operated by:
   A. the powertrain control module (PCM).
   B. the generic control module (GEM).
   C. the transmission control module (TCM).
   D. the body control module (BCM). (F.11)

4. When an automatic four-wheel-drive (A4WD) transfer case is operating in A4WD and the front driveshaft begins to turn faster than the rear driveshaft, the computer:
   A. energizes the clutch coil continually.
   B. reduces the torque to the rear wheels.
   C. increases the clutch coil duty-cycle.
   D. reduces the throttle opening. (F.12)

5. In a hydraulic clutch, the clutch fails to disengage properly when the clutch pedal is fully depressed. The cause of this problem could be:
   A. less than specified clutch pedal free play.
   B. air in the clutch hydraulic system.
   C. worn clutch facings.
   D. a scored pressure plate. (A.3)

6. When replacing a wheel stud, Technician A says that you should use a hammer to remove and install the axle shaft flange. Technician B says you will need to heat the wheel studs with a torch to remove them. Who is right?
   A. A only
   B. B only
   C. Both A and B
   D. Neither A nor B   (E.4.2)

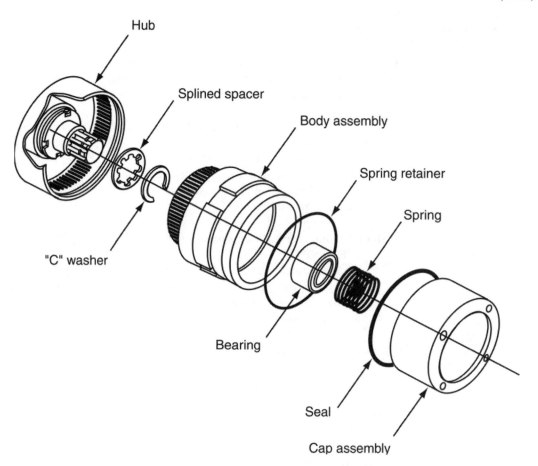

7. Technician A says automatic locking hubs should be packed with grease. Technician B says the cap on automatic locking hubs should be packed with grease. Who is right?
   A. A only
   B. B only
   C. Both A and B
   D. Neither A nor B   (F.9)

8. Excessive input shaft end play in a 4-speed transmission may cause the transmission to jump out of:
   A. first gear.
   B. second gear.
   C. fourth gear.
   D. reverse gear.   (B.12)

9. An extension housing has burrs and gouges on the mating surface. Technician A says that if they are not excessive, they can be repaired with a file. Technician B says that the mating surfaces are machined surfaces and they should only be filed with fine grit sandpaper. Who is right?
    A. A only
    B. B only
    C. Both A and B
    D. Neither A nor B   (B.14)

10. All of the following statements are true about removing and installing axle shafts **EXCEPT:**
    A. the shaft near the seal area should be inspected and repaired with sandpaper, if needed.
    B. the axle shaft seals should be replaced, not reused.
    C. the axle shaft should be stood straight up when removed.
    D. some axles require a slide hammer to remove them.   (E.4.3)

11. While discussing the major components of a hydraulic clutch linkage, Technician A says the master cylinder for the car's brakes is also used for the clutch. Technician B says the slave cylinder is connected to the clutch pedal and increases hydraulic pressure as the pedal is depressed. Who is right?
    A. A only
    B. B only
    C. Both A and B
    D. Neither A nor B   (A.3)

12. Clutch chatter could MOST Likely be caused by:
    A. excessive crankshaft end play.
    B. loose engine main bearings.
    C. a badly scored pressure plate.
    D. an improper pressure plate-to-flywheel position.   (A.5)

13. After inspecting a pilot bushing, it is found to be worn and must be replaced. Which component is LEAST Likely to be checked as a result of the find?
    A. The input shaft bearing
    B. The input shaft bearing retainer and seal
    C. The crankshaft
    D. The output shaft bushing   (A.6)

14. Technician A says if too much material is removed during flywheel resurfacing, the torsion springs on the clutch disc may contact the flywheel bolts, resulting in noise while engaging and disengaging. Technician B says if excessive material is removed when the flywheel is resurfaced, the slave cylinder rod may not have enough travel to release the clutch properly. Who is right?
    A. A only
    B. B only
    C. Both A and B
    D. Neither A nor B   (A.7)

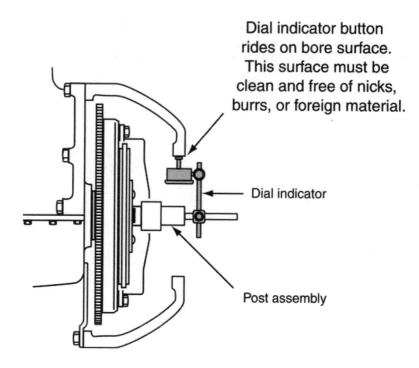

Dial indicator button
rides on bore surface.
This surface must be
clean and free of nicks,
burrs, or foreign material.

Dial indicator

Post assembly

15. Excessive misalignment between the bell housing and the engine block would MOST-Likely
cause:
   A. reduced clutch pedal free play.
   B. a growling noise when the clutch pedal is depressed.
   C. a vibration at higher speeds.
   D. clutch grabbing and chatter.   (A.8)

16. A worn pilot bearing may cause a rattling and growling noise while:
   A. the engine is idling and the clutch pedal is fully depressed.
   B. decelerating in high gear with the clutch pedal released.
   C. accelerating in low gear with the clutch pedal released.
   D. the engine is idling in neutral with the clutch pedal released.   (A.1)

17. Technician A says sagged transmission mounts may cause improper driveshaft angles on a
rear-wheel drive car. Technician B says improper driveshaft angles may cause a constant-speed
vibration when the vehicle is accelerated and decelerated. Who is right?
   A. A only
   B. B only
   C. Both A and B
   D. Neither A nor B   (A.10)

18. All of the following problems may cause a vibration on a four-wheel drive vehicle that is more
noticeable when changing throttle position **EXCEPT:**
   A. worn U-joints.
   B. worn front-drive axle joints.
   C. incorrect driveshaft angles.
   D. worn driveshaft slip joint splines.   (F.7)

19. The counter/cluster gear shaft and needle bearings are pitted and scored. Technician A says the transmission may have a growling noise with the engine idling, transmission in neutral, and the clutch pedal out. Technician B says the transmission may have a growling noise while driving in any gear. Who is right?
   A. A only
   B. B only
   C. Both A and B
   D. Neither A nor B   (B.1)

20. Technician A says that repeated extension housing seal failures may be caused by a worn extension housing bushing. Technician B says that a worn countershaft bearing may cause an extension housing bushing to wear prematurely. Who is right?
   A. A only
   B. B only
   C. Both A and B
   D. Neither A nor B   (B.3)

21. A five-speed manual transaxle has a growling and rattling noise in third gear only. The cause of this problem could be:
   A. worn, chipped teeth on the third-speed gear on the input shaft.
   B. worn dog teeth on the third-speed gear on the input shaft.
   C. worn dog teeth on the third-speed synchronizer blocker ring.
   D. worn threads in the cone area of the third-speed blocker ring.   (C.7)

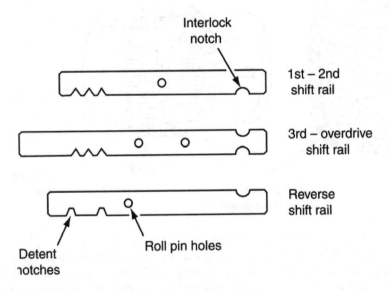

22. A bent shift rail will MOST-Likely cause:
   A. the transmission to jump out of gear.
   B. hard shifting in some gears.
   C. gear clash during some shifts.
   D. gear noise in some gears.   (B.6)

23. Case/carrier bearing preload is correct in a shim-type differential, but backlash is excessive. To correct this, you should:
    A. remove shims from the right side and add shims to the left.
    B. add an equal number of shims to both sides.
    C. remove shims from the left side and add shims to the right.
    D. remove an equal number of shims from both sides.   (C.17)

24. A manual transmission, when in neutral, has a growling noise with the engine idling and the clutch released. The noise disappears when the clutch pedal is depressed. The cause of this noise could be a worn:
    A. pilot bearing in the crankshaft.
    B. input shaft and pilot bearing contact area.
    C. input shaft bearing.
    D. mainshaft ball bearing.   (B.7)

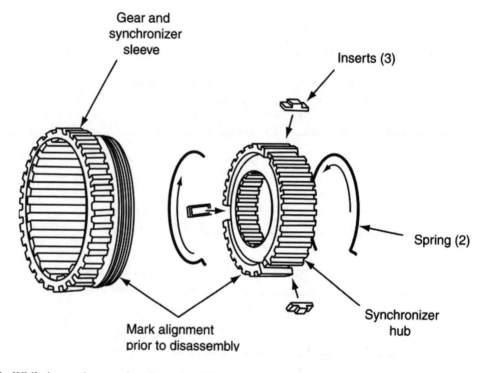

25. While inspecting synchronizer assemblies on a manual transmission:
    A. the dog teeth tips on the blocking rings should be flat with smooth surfaces.
    B. the threads in the cone area of the blocking rings should be sharp and not dull.
    C. the clearance is not important between the blocking ring and the matching gear's dog teeth.
    D. the sleeve should fit snugly on the hub and offer a certain amount of resistance to movement.   (B.9)

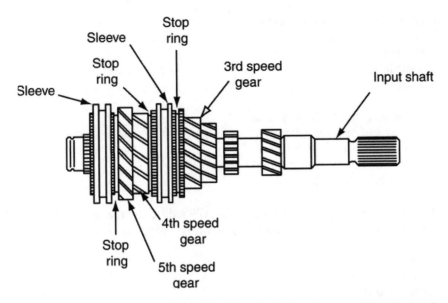

26. The clearance on the fourth-speed gear blocking ring is less than specified. Technician A says this may result in noise while driving in fourth gear. Technician B says this problem may cause hard shifting into fourth gear. Who is right?
    A. A only
    B. B only
    C. Both A and B
    D. Neither A nor B   (B.9)

27. A transfer case is considered full with fluid when:
    A. fluid comes out of the front input when the yoke is removed.
    B. fluid is at the halfway point of the sight glass window.
    C. fluid just starts to come back out of the fill hole when the plug is removed.
    D. at least 4 inches (102 mm) of fluid level is found when a piece of mechanic's wire is inserted into the fill hole.   (E.3.2)

28. Excessive end play in the counter gear could cause damage to all of the following **EXCEPT:**
    A. the transmission case.
    B. the gear teeth.
    C. the counter shaft.
    D. the release bearing.   (B.10)

29. A transaxle shifts normally into all forward gears, but it will not shift into reverse gear; there is no evidence of noise while attempting this shift. Technician A says the reverse shifter fork may be broken. Technician B says the reverse idler gear teeth may be worn. Who is right?
    A. A only
    B. B only
    C. Both A and B
    D. Neither A nor B   (C.10)

30. Which of the following is LEAST Likely to be replaced when replacing the extension?
    A. The output shaft seal
    B. The tail housing gasket
    C. The speedometer O-ring
    D. The counter gear shaft   (B.14)

31. The speedometer driving gear is located on the:
    A. vehicle speed sensor.
    B. main shaft.
    C. output shaft.
    D. axle shaft.   (B.15)

32. A remote vent on a differential is used to:
    A. increase pressure in the differential.
    B. keep moisture out of the differential.
    C. keep lubricant from coming out of the differential.
    D. add lubricant to the differential.   (F.10)

33. Technician A says compared to hypoid gear oil, ATF reduces friction and improves fuel
    economy. Technician B says the thicker the gear oil, the lower the viscosity number. Who is
    right?
    A. A only
    B. B only
    C. Both A and B
    D. Neither A nor B   (B.17)

34. A fully synchronized four-speed manual transaxle experiences gear clash in all forward gears
    and reverse. The cause of this problem could be:
    A. a worn fourth-gear synchronizer.
    B. the clutch disc sticking on the input shaft.
    C. a worn blocker ring.
    D. a worn ¾ shifter fork.   (C.1)

35. Technician A says an improper shift linkage adjustment may cause hard transaxle shifting.
    Technician B says an improper shift linkage adjustment may cause the transaxle to stick in
    gear. Who is right?
    A. A only
    B. B only
    C. Both A and B
    D. Neither A nor B   (C.2)

36. A transaxle has had repeated drive axle seal leakage and replacement. Technician A says this
    problem may be caused by a plugged transaxle vent. Technician B says this problem may be
    caused by a worn outer drive axle joint. Who is right?
    A. A only
    B. B only
    C. Both A and B
    D. Neither A nor B   (C.3)

37. Technician A says a misaligned engine and transaxle cradle may cause drive axle vibrations.
    Technician B says a misaligned engine and transaxle cradle may cause improper front suspension
    angles. Who is right?
    A. A only
    B. B only
    C. Both A and B
    D. Neither A nor B   (C.4)

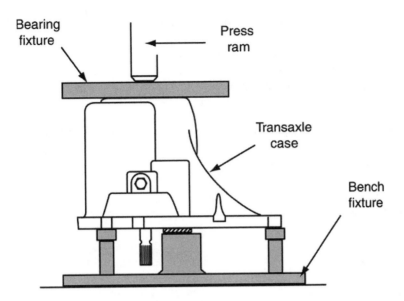

38. After removal of a transaxle, all of the following components will require opening the transaxle case before removal **EXCEPT:**
    A. the input shaft.
    B. the output shaft.
    C. the pilot bearing.
    D. the differential bearings.   (C.5)

39. Technician A says in four-wheel drive (4WD) low, the power flow in the transfer case goes from the input shaft through the sun gear and planetary carrier to provide gear reduction. Technician B says in 4WD low, the annulus gear in the planetary gear is rotating counterclockwise. Who is right?
    A. A only
    B. B only
    C. Both A and B
    D. Neither A nor B   (F.2)

40. A four-speed manual transaxle jumps out of third gear. Technician A says the shift rail detent spring tension on the 3-4 shift rail may be weak. Technician B says there may be excessive wear on the fourth-speed gear dog teeth. Who is right?
    A. A only
    B. B only
    C. Both A and B
    D. Neither A nor B   (C.6)

41. A broken detent spring could cause any of the following problems **EXCEPT:**
    A. the transmission to jump out of gear.
    B. harsh shifting.
    C. lock up between two gears.
    D. a growling noise.   (C.6)

42. The second-speed gear dog teeth and blocking ring teeth are badly worn. This problem may cause:
    A. a growling noise while driving in second gear.
    B. a vibration while accelerating in second gear.
    C. hard shifting in second and third gear.
    D. the transaxle to jump out of second gear.   (C.8)

43. Technician A says synchronizer hubs on a manual transaxle are reversible on the shaft on which they are mounted. Technician B says synchronizer sleeves on a manual transaxle are reversible on their matching hub. Who is right?
    A. A only
    B. B only
    C. Both A and B
    D. Neither A nor B   (C.9)

44. In a manual transmission, the reverse idler gear only is in mesh with what gear?
    A. First gear
    B. Fifth gear
    C. Third and fourth gears
    D. Reverse gear   (B.11)

45. In some transaxles, the speedometer drive gear is mounted on:
    A. the input shaft.
    B. the transfer gear.
    C. the output shaft.
    D. the drive axle inner hub.   (C.12)

46. Technician A says incorrect fluid type can cause transmission leakage. Technician B says incorrect fluid type could affect fuel economy. Who is correct?
    A. A only
    B. B only
    C. Both A and B
    D. Neither A nor B   (B.17 and B.18)

47. Technician A says vehicle speed sensors can be tested with an ammeter. Technician B says when you check a vehicle speed sensor signal, you should use an oscilloscope while the vehicle is being driven. Who is right?
    A. A only
    B. B only
    C. Both A and B
    D. Neither A nor B   (C.13)

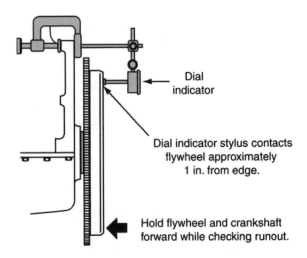

Dial indicator

Dial indicator stylus contacts flywheel approximately 1 in. from edge.

Hold flywheel and crankshaft forward while checking runout.

48. With the dial indicator positioned as shown in the figure, the measurement being performed is:
    A. crankshaft end play.
    B. crankshaft warpage.
    C. rear main bearing wear.
    D. rear engine block alignment.   (A.9)

49. A constant whining noise is coming from the differential of a vehicle. Which of the following could be the cause?
    A. The preload and backlash are not set properly.
    B. The wrong differential lube was used.
    C. The side gears are damaged.
    D. The spider gears are damaged.   (C.14)

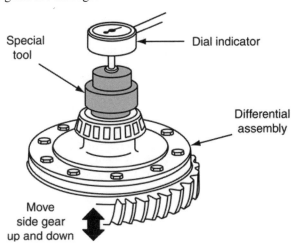

Special tool

Dial indicator

Differential assembly

Move side gear up and down

50. While determining the proper differential side gear thrust washer thickness, on a FWD final drive:
    A. measure the end play only on one side gear to calculate the side gear spacer washer thickness.
    B. measure the side gear end play with the thrust washers behind the gears.
    C. the correct thickness of the side gear thrust washer provides the specified side gear end play.
    D. the correct thickness of the side gear thrust washer provides a slight side gear preload.   (C.16)

51. Technician A says that stripped threads in a transaxle case can be repaired with a helicoil. Technician B says that not all damaged threads in a transaxle should be repaired. Who is right?
    A. A only
    B. B only
    C. Both A and B
    D. Neither A nor B   (C.11)

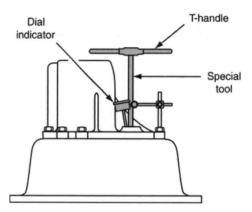

52. When taking the differential end play measurement shown in the figure,:
    A. a new bearing cup is installed with a shim in the clutch bell housing side of the transaxle case.
    B. a medium load should be applied to the differential in the upward direction.
    C. the proper shim thickness is equal to the differential end play recorded on the dial indicator.
    D. the bolts between the transaxle case halves must be tightened to one-half the specified torque.   (C.18)

53. The needle bearings between the output shaft and the output shaft gears are scored and blue from overheating. Technician A says the transaxle may have been filled with the wrong lubricant. Technician B says the projection is broken off the oil feeder behind the front output shaft bearing. Who is right?
    A. A only
    B. B only
    C. Both A and B
    D. Neither A nor B   (C.19)

54. Technician A says a worn U-joint may cause a squeaking noise. Technician B says a heavy vibration that only occurs during acceleration may be caused by a worn centering ball and socket on a double Cardan U-joint. Who is right?
    A. A only
    B. B only
    C. Both A and B
    D. Neither A nor B   (D.1)

55. Technician A says the pinion bearings should be lubricated when the pinion turning torque is measured. Technician B says the pinion nut may be loosened to obtain the specified turning torque. Who is right?
    A. A only
    B. B only
    C. Both A and B
    D. Neither A nor B   (E.1.5)

56. A clicking noise is heard on a front-wheel drive vehicle while turning a corner. The cause of this problem could be a bad:
    A. front axle shaft.
    B. CV inner joint.
    C. torsional damper.
    D. outer CV joint.   (D.1)

57. A front-wheel drive car has a clicking noise while turning. Technician A says this may be caused by a worn outer-drive axle joint. Technician B says this may be caused by a front wheel bearing. Who is right?
    A. A only
    B. B only
    C. Both A and B
    D. Neither A nor B   (D.2)

58. Technician A says that if the synchronizer sleeve does not slide smoothly over the blocker ring and gear teeth, shifting may not occur. Technician B says the synchronizer hub and sleeve must be marked in relation to each other before disassembly. Who is right?
    A. A only
    B. B only
    C. Both A and B
    D. Neither A nor B   (B.5)

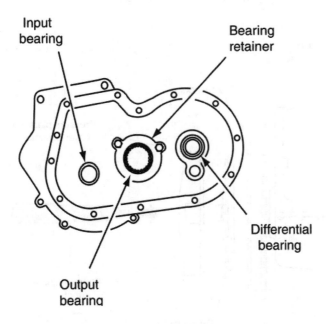

59. To adjust the differential side bearing preload in the transaxle shown in the figure, Technician A says preload is adjusted by rotating a threaded adjuster on each side of the differential bearings. Technician B says the preload is automatically adjusted when the case halves are reassembled. Who is right?
    A. A only
    B. B only
    C. Both A and B
    D. Neither A nor B   (C.15)

60. To measure axle shaft end play:
    A. the differential cover must be removed.
    B. the vehicle needs to be in neutral.
    C. the brake drum needs to be removed.
    D. the axle shaft needs to be removed from the axle housing.   (E.4.4)

61. A center shaft support bearing:
    A. should be greased like a universal joint.
    B. is usually a sealed bearing and is maintenance-free.
    C. is used on all rear-wheel drive vehicles.
    D. is part of the driveshaft and cannot be replaced separately.   (D.3)

62. When replacing a differential, Technician A says that the differential carrier bearings can be
    used with the new differential. Technician B says that if the differential bearings are replaced,
    the bearing races must be replaced. Who is right?
    A. A only
    B. B only
    C. Both A and B
    D. Neither A nor B   (E.2.2)

63. Technician A says the dial indicator should be positioned near the front of the driveshaft to
    measure driveshaft runout. Technician B says if the driveshaft runout is excessive, the driveshaft
    may be straightened in a hydraulic press. Who is right?
    A. A only
    B. B only
    C. Both A and B
    D. Neither A nor B   (D.5)

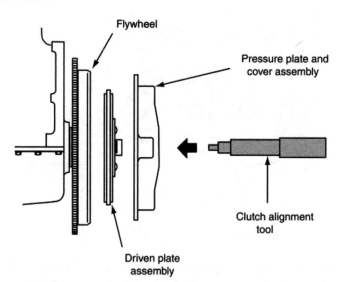

64. During manual transmission removal and replacement:
    A. the driveshaft may be installed in any position on the differential pinion gear flange.
    B. the transmission weight may be supported by the input shaft in the clutch disc hub.
    C. the engine support fixture should be installed after the transmission-to-engine bolts are
       loosened.
    D. the clutch disc must be aligned with an aligning tool before transmission installation. (B.4)

65. While a vehicle is moving straight ahead, the differential produces a whining noise. Technician A says that the differential side gears may be damaged. Technician B says that the ring and pinion gear adjustments may be incorrect. Who is right?
    A. A only
    B. B only
    C. Both A and B
    D. Neither A nor B   (E.1.1)

66. Technician A says that spider gears are used when the vehicle is cornering. Technician B says the individual spider gears rotate at different speeds when the vehicle is going straight. Who is right?
    A. A only
    B. B only
    C. Both A and B
    D. Neither A nor B   (E.1.1)

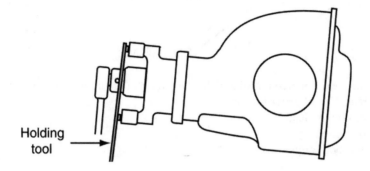

Holding tool

67. Technician A says that insufficient pinion nut torque may cause a clunking noise during acceleration or deceleration. Technician B says that insufficient pinion nut torque may cause a growling noise with the vehicle in motion. Who is right?
    A. A only
    B. B only
    C. Both A and B
    D. Neither A nor B   (E.1.2)

68. A transmission has a growling noise in first gear only. The cause of the problem could be:
    A. a worn first-gear blocker ring.
    B. a worn first-gear synchronizer ring.
    C. a chipped or worn first-speed gear tooth.
    D. worn main shaft bearings.   (B.8)

69. A front axle on a four-wheel drive (4WD) vehicle is leaking oil. Technician A says that the axle vent may be plugged. Technician B says that the axle may be overfilled with lubricant. Who is right?
    A. A only
    B. B only
    C. Both A and B
    D. Neither A nor B   (E.2.6)

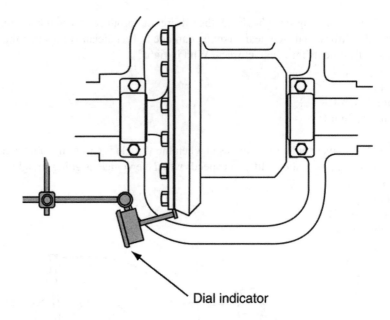

Dial indicator

70. Excessive ring gear runout on the dial indicator shown in the figure may be caused by excessive:
    A. differential case/carrier runout.
    B. side bearing preload.
    C. side gear end play.
    D. ring gear bolt torque.   (E.1.3)

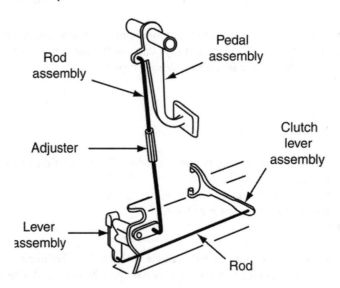

71. As shown in the figure, lack of clutch pedal free play may cause:
    A. hard shifting.
    B. incomplete clutch release.
    C. transaxle gear damage.
    D. clutch slipping.   (A.2)

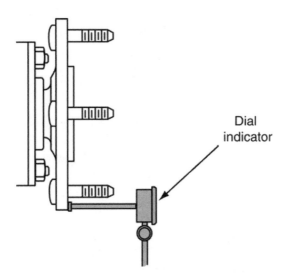

72. The cause of excessive runout on the dial indicator shown in the figure could be:
    A. incorrect differential bearing preload.
    B. a bent axle shaft.
    C. incorrect pinion depth.
    D. worn limited slip clutch discs.   (E.4.4)

73. Technician A says the collapsible pinion shaft spacer may be reused if the differential is
    disassembled and overhauled. Technician B says that after proper pinion bearing preload is set,
    pinion depth is adjusted by backing off the pinion flange nut. Who is right?
    A. A only
    B. B only
    C. Both A and B
    D. Neither A nor B   (E.1.4)

74. Technician A says the clutch pedal free play adjustment sets the distance between the release
    bearing and the pressure plate fingers. Technician B says a worn release bearing creates the most
    noise when the clutch pedal is released. Who is right?
    A. A only
    B. B only
    C. Both A and B
    D. Neither A nor B   (A.4)

75. When adjusting the pinion depth on the ring gear, Technician A says to replace the collapsible
    spacer. Technician B says that pinion depth can also be adjusted by installing a selective pinion
    bearing race. Who is right?
    A. A only
    B. B only
    C. Both A and B
    D. Neither A nor B   (E.1.5)

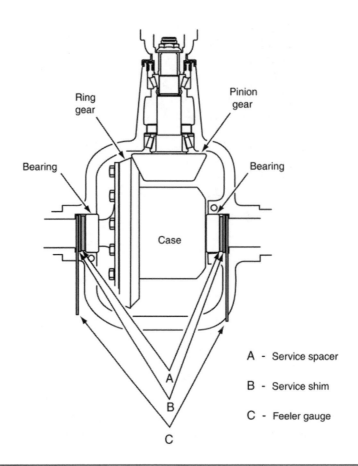

EXAMPLE		
	Ring gear side Combined total of:	Opposite side Combined total of:
.250"	Service spacer (A) Service shim (B) Feeler gauge (C)	Service spacer (A) Service shim (B) Feeler gauge (C) .265"
-.010" ‾‾‾‾‾ .240"	To maintain proper backlash (.005"–.008") ring gear is moved away from the pinion by subtracting .010" shim from ring gear side and adding .010" to the other side	+.010" ‾‾‾‾‾ .275"
+.004"	To obtain proper preload on side bearings, add .004" shim to each side	+.004"
.244"	Shim dimension required for ring gear side	Shim dimension required for opposite side .279"

76. Technician A says if the ring tooth contact pattern indicates pinion tooth contact on the toe of the pinion gear, the pinion gear should be moved toward the ring gear. Technician B says if the pinion gear teeth have low flank contact on the ring gear teeth, the pinion gear should be moved toward the ring gear. Who is right?
    A. A only
    B. B only
    C. Both A and B
    D. Neither A nor B

    (E.1.6)

77. After inspecting the ring gear, a toe pattern is found. Which of the following is the MOST-Likely cause?
    A. A worn axle bearing
    B. A worn collapsible pinion spacer
    C. Worn spider gears
    D. A worn differential bearing (E.1.8)

78. When diagnosing a rear-wheel drive differential that vibrates only while turning a corner, Technician A says the pinion bearings may be worn, and the pinion bearing preload is less than specified. Technician B says the bearing surfaces between the side gears and the differential case may be damaged. Who is right?
    A. A only
    B. B only
    C. Both A and B
    D. Neither A nor B (E.2.1)

79. When removing a transfer case from the vehicle, which of the following components listed is LEAST Likely to be disconnected or removed?
    A. The front driveshaft
    B. The transmission
    C. The rear driveshaft
    D. The linkage (F.3)

80. Which of these is the MOST-Likely cause of clutch chatter?
    A. A worn, rough clutch release bearing
    B. A worn, rough pilot bearing
    C. Excessive input shaft end play
    D. Weak clutch plate torsional springs (A.1)

81. To measure differential case/carrier runout, which of the following tools would be used?
    A. Torque wrench
    B. Dial indicator
    C. Feeler gauge
    D. Micrometer (E.2.5)

82. Too much driveline angle would MOST Likely cause:
    A. binding universal joints.
    B. a loud humming noise.
    C. pinion gear damage.
    D. damage to the transmission mount. (D.6)

83. A limited slip differential chatters while cornering. Technician A says the differential may be filled with the wrong lubricant. Technician B says friction and steel plates may be worn and burned. Who is right?
    A. A only
    B. B only
    C. Both A and B
    D. Neither A nor B (E.3.1)

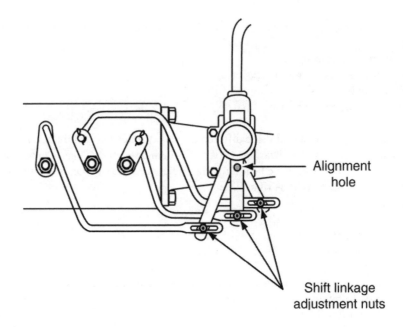

84. The shift lever adjustment is usually performed with the transmission in:
    A. neutral.
    B. first gear.
    C. second gear.
    D. reverse gear.   (B.2)

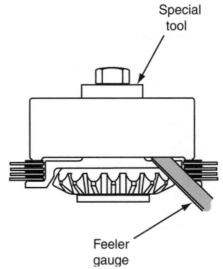

85. The measurement in the figure shown determines the proper:
    A. friction plate thickness.
    B. steel plate thickness.
    C. shim thickness.
    D. preload spring tension.   (E.3.3)

86. An all-wheel drive vehicle has a vibration that is more noticeable while cornering. Technician A says the U-joints may be worn. Technician B says the outboard front axle joints may be worn. Who is right?
    A. A only
    B. B only
    C. Both A and B
    D. Neither A nor B   (F.1)

87. While discussing backup lamp switches, Technician A says that a backup lamp switch is located in the transmission and is normally open. Technician B says that a backup lamp switch can have power to the switch in RUN and is closed when the vehicle is shifted into reverse. Who is right?
    A. A only
    B. B only
    C. Both A and B
    D. Neither A nor B   (B.16)

88. Technician A says a plugged transfer case vent may cause seal leakage. Technician B says that a remote transfer case vent helps to prevent moisture from entering the transfer case when driving through water. Who is right?
    A. A only
    B. B only
    C. Both A and B
    D. Neither A nor B   (F.4)

89. Premature wear in the extension housing bushing will MOST Likely be caused by:
    A. worn speedometer drive and driven gears.
    B. uneven mating surfaces on the extension housing.
    C. a plugged transmission vent opening.
    D. excessive transmission main shaft end play.   (B.13)

6 Additional Test Questions for Practice

Additional Test Questions

Please note the letter and number in parentheses following each question. They match the task in Section 4 that discusses the relevant subject matter. You may want to refer to the overview using the cross-referencing key to help with questions posing problems for you.

1. If an axle shaft has seal surface damage, all of the following should be inspected **EXCEPT:**
   A. the axle seal.
   B. the axle bearing.
   C. the rear axle housing.
   D. the brake drum. (E.4.3)

2. All of the following can cause axle shaft seal leakage **EXCEPT:**
   A. worn axle shaft bearings.
   B. plugged axle vent.
   C. scored axle shafts.
   D. heat from the rear brakes. (E.4.3)

3. When an axle shaft is removed from the differential, it should be:
   A. stored standing straight up.
   B. laid flat.
   C. rolled on the ground to check for runout.
   D. left under the vehicle. (E.4.3)

4. An input shaft is supported by and uses:
   A. bronze bushings.
   B. needle bearings.
   C. a sealed bearing.
   D. ball bearings. (B.7)

5. A vacuum-shifted four-wheel drive (4WD) system does not shift into 4WD. Technician A says the engine vacuum may be low. Technician B says the vacuum motor at the front axle may be the problem. Who is right?
   A. A only
   B. B only
   C. Both A and B
   D. Neither A nor B (F.1)

6. After inspecting a transmission mount and finding it oil soaked, Technician A says it should be replaced. Technician B says some mounts can be cleaned and reinstalled when oil soaked. Who is right?
   A. A only
   B. B only
   C. Both A and B
   D. Neither A nor B (A.10)

7. An electronic transfer case does not engage. Which of the following would be the LEAST-Likely cause?
    A. A bad electronic shift motor
    B. A blown fuse
    C. The 4WD engage switch
    D. A rusted linkage   (F.11)

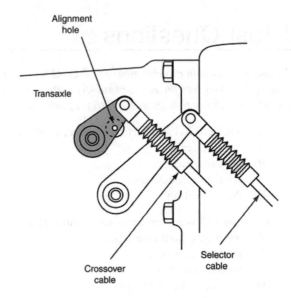

8. In the figure shown, the transaxle is in the neutral position to allow the technician to:
    A. replace the shift cables.
    B. adjust the shift cables.
    C. modify the shift cables.
    D. repair the transaxle case.   (C.2)

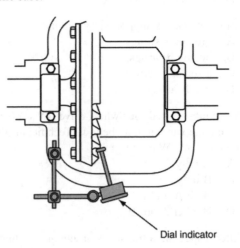

9. In the figure shown, a technician is measuring:
    A. ring gear runout.
    B. ring gear backlash.
    C. pinion gear backlash.
    D. bearing preload.   (E.1.7)

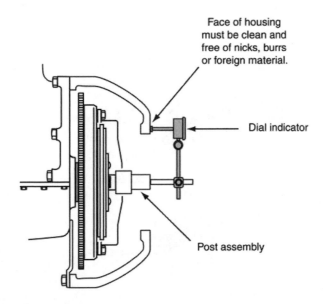

Face of housing
must be clean and
free of nicks, burrs
or foreign material.

Dial indicator

Post assembly

10. To correct excessive runout, as shown on the dial indicator:
    A. replace the engine mounts.
    B. adjust eccentric bell housing dowels.
    C. install bell housing shims.
    D. replace the clutch disc.   (A.10)

11. A limited-slip differential is not working properly. Which of the following would be the
    LEAST-Likely cause?
    A. Worn friction plates
    B. Weak spring tension
    C. Stripped teeth on the friction plates
    D. Wrong fluid in the differential   (E.3.1)

12. Technician A says that the blocking ring with the sharp edges is good. Technician B says that the
    sharp edges will prevent the transmission from jumping out of gear. Who is right?
    A. A only
    B. B only
    C. Both A and B
    D. Neither A nor B   (B.9)

13. Technician A says that when filling a transfer case, fluid level can be checked by looking in the
    sight glass. Technician B says to fill the transfer case until it begins to come out of the fill
    vent. Who is right?
    A. A only
    B. B only
    C. Both A and B
    D. Neither A nor B   (E.3.2)

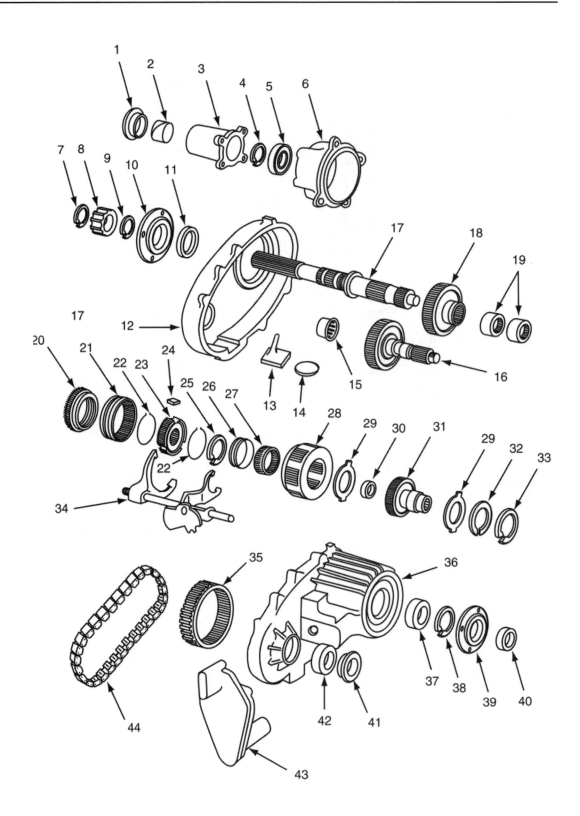

1. Rear output shaft seal	23. Sychronizer hub
2. Extension housing bushing	24. Sychronizer struts (3)
3. Extension housing	25. Snap ring
4. Retainer	26. Range shift hub
5. Rear output shaft bearing	27. Range gear
6. Pump retaininer hiusing	28. Planetary carrier
7. Tone wheel retainer	29. Thrust washer (2)
8. Tone wheel	30. Mainshaft pilot bearing
9. Tone wheel retainer	31. Input gear
10. Oil pump	32. Carrier lock ring
11. Oil pump seal	33. Snap ring
12. Rear case half	34. Shifting fork mechanism
13. Pump pick-up screen	35. Annulus gear
14. Magnet	36. Front case half
15. Front output rear bearing	37. Input bearing
16. Front output (driven) shaft	38. Snap ring
17. Main shaft	39. Input bearing retainer
18. Drive sprocket	40. Input bearing seal
19. Drive sprocket bearings	41. Front output shaft seal
20. Main drive sychronizer ring	42. Front output shaft bearing
21. Sychronizer sleeve	43. Encoder motor
22. Sychronizer strut spring (2)	44. Drive chain

14. All of the following are true about transfer case inspection and assembly **EXCEPT:**
    A. the transfer case chain should be inspected for stretching and looseness.
    B. thrust washer thickness should be measured to check for wear or for select fit.
    C. specification measurements should have been taken during disassembly.
    D. all parts should be cleaned before installation and assembled dry.   (F.2)

15. Technician A says that driveshaft runout should be checked with a digital micrometer set in the middle of the driveshaft. Technician B says that a dial indicator should be set at the differential end of the driveshaft to check runout. Who is right?
    A. A only
    B. B only
    C. Both A and B
    D. Neither A nor B   (D.5)

16. Technician A says improper fluid type can cause manual transmission shift problems. Technician B says improper fluid type can cause accelerated blocking ring wear. Who is correct?
    A. A only
    B. B only
    C. Both A and B
    D. Neither A nor B   (B.17 and B.18)

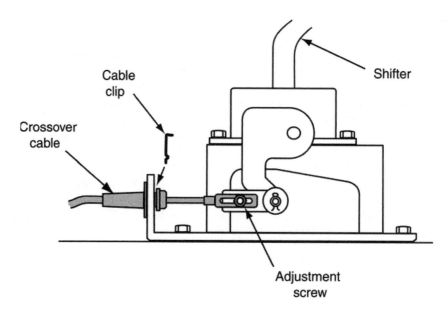

17. In the figure, Technician A says to check the selector cable if the vehicle has a sloppy shifter. Technician B says to check all external linkage bushings and grommets. Who is right?
    A. A only
    B. B only
    C. Both A and B
    D. Neither A nor B   (C.2)

18. During a transmission inspection, a damaged input bearing is found. Which of the following is also likely to be damaged?
    A. The bearing retainer and seal
    B. The counter shaft bearing
    C. Second gear
    D. The counter gear   (B.7)

19. Technician A says that a worn blocker ring will cause the transaxle to have gear clash. Technician B says that a worn blocker ring will have sharp ridges in the cone area from slipping on the gear cone area. Who is right?
    A. A only
    B. B only
    C. Both A and B
    D. Neither A nor B   (C.9)

20. Bearing preload is:
    A. to pack a bearing before installation.
    B. the amount of pressure applied to a bearing while the transmission is under load.
    C. the amount of pressure applied to a bearing upon assembly of the transmission.
    D. not adjustable.   (B.13)

21. To measure driveshaft runout, you should do all of the following **EXCEPT:**
    A. place the vehicle transmission in neutral.
    B. use a magnetic-base dial indicator.
    C. roll the driveshaft on a flat surface to check for damage.
    D. clean the driveshaft surface for an accurate runout check.   (D.5)

22. While inspecting a reverse idler gear from a manual transmission, Technician A says that you should check the center bore for a smooth mar-free surface. Technician B says that the reverse idler gear is splined in the center bore and that the splines should be checked for excessive wear or damage. Who is right?
    A. A only
    B. B only
    C. Both A and B
    D. Neither A nor B   (B.11)

23. After inspection of the reverse idler gear, Technician A says that since one tooth on the counter/cluster gear is chipped, the entire gear must be replaced. Technician B says a reverse idler gear with a chipped tooth does not have to be replaced because it is only used in reverse. Who is right?
    A. A only
    B. B only
    C. Both A and B
    D. Neither A nor B   (B.11)

24. The speedometer needle bounces when the vehicle is moving. Technician A says that the gear may be worn on the output shaft. Technician B says there could be a bad cable or speedometer head. Who is right?
    A. A only
    B. B only
    C. Both A and B
    D. Neither A nor B   (B.13)

25. A bearing-type noise begins to come from the clutch and transmission area of a vehicle just as the clutch is almost completely disengaged. There is no noise when the clutch pedal is initially depressed. Technician A says that the clutch pilot bearing may be worn out. Technician B says that the transmission input shaft bearing may be faulty. Who is right?
    A. A only
    B. B only
    C. Both A and B
    D. Neither A nor B   (A.1)

26. A transmission will not engage in reverse when shifted. Which of these is the LEAST-Likely cause?
    A. A bent linkage
    B. A broken shift fork
    C. A misadjusted linkage
    D. Damaged teeth on the gears   (B.6)

27. A clutch makes a loud "chirping" noise when the clutch pedal is depressed with the engine running. Technician A says that the release bearing is worn and must be replaced. Technician B says that the input shaft bearing is worn and must be replaced. Who is right?
    A. A only
    B. B only
    C. Both A and B
    D. Neither A nor B   (A.1)

28. Technician A says that clutch chatter could be caused by an uneven flywheel. Technician B says that oil on the clutch disc can cause clutch chatter. Who is right?
    A. A only
    B. B only
    C. Both A and B
    D. Neither A nor B   (A.1)

29. The backup lights are staying on in all gears. What is the LEAST-Likely cause?
    A. A short in the backup light switch
    B. The linkage out of adjustment
    C. A short in the wiring harness
    D. A bad brake light switch   (B.16)

30. Technician A says that a transmission that is hard to shift may have a problem with the linkage not being lubricated. Technician B says that this problem may be caused by too strong of a pressure plate installed in the vehicle's clutch system. Who is right?
    A. A only
    B. B only
    C. Both A and B
    D. Neither A nor B   (B.1)

31. When checking a diagnostic trouble code (DTC) that has to do with the output shaft speed sensor, which of the following should NOT be done next after reading the code?
    A. Install a new sensor
    B. Inspect the harness for continuity
    C. Check the service manual
    D. Inspect the connector   (B.16)

32. When reassembling a transmission shaft, gear and synchronizer end play are set by using:
    A. a new main shaft.
    B. all new bearings.
    C. select-fit shims and thrust washers.
    D. all new snap rings.   (B.12)

33. While disassembling a manual transmission for an overhaul, all of the following will need to be done **EXCEPT:**
    A. take and record measurements.
    B. clean all the parts with cleaning solvent.
    C. pay attention to the condition of the parts that are removed.
    D. keep the friction discs in the correct order.   (B.5)

34. Technician A says that when a wheel bearing is replaced, the bearing race should also be replaced. Technician B says the automatic locking hubs and the caps should be packed with grease. Who is right?
    A. A only
    B. B only
    C. Both A and B
    D. Neither A nor B   (F.9)

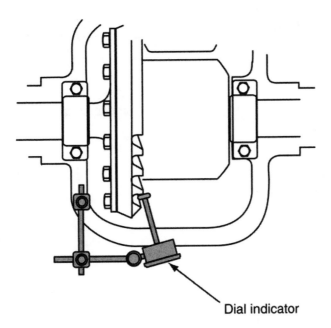

Dial indicator

35. In the figure shown, the ring gear backlash and side play are zero. Right- and left-side bearing adjustment nuts are determined while facing the differential from the rear. To obtain the proper ring gear backlash:
    A. tighten the right and left-side bearing adjusters.
    B. loosen the left-side bearing adjuster.
    C. loosen the left-side bearing adjuster and tighten the right-side bearing adjuster.
    D. loosen the right-side bearing adjuster and tighten the left-side bearing adjuster.

                                                                        (E.1.7 and E.2.5)

36. While discussing the main shaft third gear dog clutch teeth, Technician A says the teeth should be rounded. Technician B says the teeth should have a sharp, beveled edge. Who is right?
    A. A only
    B. B only
    C. Both A and B
    D. Neither A nor B   (B.9)

37. Most late-model manual transmissions are filled with:
    A. SAE 90 gear oil.
    B. power steering fluid.
    C. automatic transmission fluid or motor oil.
    D. a synthetic lubricant with a Teflon additive.   (B.17)

38. A manual transmission jumps out of second gear. Technician A says there may be excessive end play between the second-speed gear and its matching synchronizer. Technician B says the detent springs on the shift rail may be weak. Who is right?
    A. A only
    B. B only
    C. Both A and B
    D. Neither A nor B   (B.1)

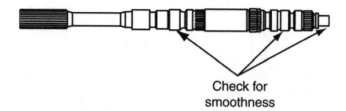

Check for
smoothness

39. The figure shows a main shaft; the areas pointed out are:
    A. bearing surfaces.
    B. gear journals.
    C. oil journal locations.
    D. synchronizer mounting locations.   (B.8)

40. A vehicle with a manual transaxle makes noise that is loudest during turns. What would be
    the MOST-Likely cause?
    A. The clutch throwout bearing
    B. Ring gear and pinion teeth
    C. Differential side bearings
    D. The wrong lubricant in the differential   (C.1 and C.14)

41. Technician A says that transmissions with internal linkage have no internal adjustment.
    Technician B says that only transmissions with external linkage can be adjusted. Who is right?
    A. A only
    B. B only
    C. Both A and B
    D. Neither A nor B   (B.2)

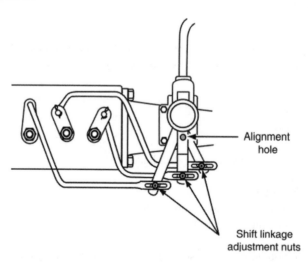

Alignment
hole

Shift linkage
adjustment nuts

42. When replacing transmission shift levers as shown in the figure, Technician A says that all lever
    bushings should be replaced. Technician B says that only the bushings that are worn should
    be replaced. Who is right?
    A. A only
    B. B only
    C. Both A and B
    D. Neither A nor B   (B.2)

43. Technician A says constant-running release bearings are used with hydraulically controlled clutches. Technician B says release bearings move away from the pressure plate to disengage the clutch. Who is right?
    A. A only
    B. B only
    C. Both A and B
    D. Neither A nor B   (A.4)

44. Technician A says that when removing a manual transmission, always check for a rear main engine seal leak. Technician B says that the engine does not have to be supported when the transmission is removed from the vehicle. Who is right?
    A. A only
    B. B only
    C. Both A and B
    D. Neither A nor B   (B.4)

45. A transmission mount saturated with oil should be:
    A. cleaned and reused.
    B. inspected for cracks.
    C. used for the rubber in the mount.
    D. discarded and replaced.   (B.4)

46. Technician A says that the input shaft end play does not need to be checked before disassembling a transaxle. Technician B says to always rotate the input shaft to check turning effort before disassembling the transaxle. Who is right?
    A. A only
    B. B only
    C. Both A and B
    D. Neither A nor B   (C.5)

47. All of the following are true about removing a differential assembly **EXCEPT:**
    A. the axle shafts must be removed.
    B. the bearing caps should be marked to the housing.
    C. the bearing races and shim pack should not be mixed up.
    D. the pinion gear always stays in the axle housing.   (E.2.2)

48. Which of the following is the best way to check component wear when assembling a transfer case?
    A. Prelube the components
    B. Clean components thoroughly
    C. Clean all sealing surfaces
    D. Measure clearances   (F.2)

49. While discussing transaxle case mating surfaces, Technician A says that the mating surfaces do not have to be machined to precise specifications because the gasket or sealer will stop any leaks. Technician B says that a transaxle that has two halves to its case has a machined mating surface, but still requires a gasket or a sealant to prevent leakage. Who is right?
    A. A only
    B. B only
    C. Both A and B
    D. Neither A nor B   (C.3)

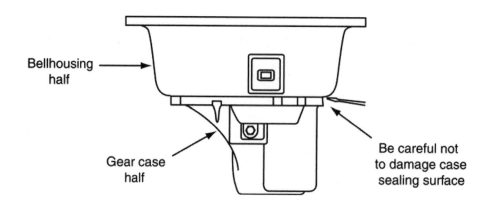

50. While discussing manual transaxles, Technician A says that most transaxle cases are sealed with room temperature vulcanizing (RTV) sealant. Technician B says that some transaxle cases have paper gaskets. Who is right?
    A. A only
    B. B only
    C. Both A and B
    D. Neither A nor B   (C.3)

51. Technician A says that a worn pilot bearing or bushing is a common result of a misalignment condition. Technician B says that a bent transmission input shaft or a bent clutch disc may cause a misalignment condition. Who is right?
    A. A only
    B. B only
    C. Both A and B
    D. Neither A nor B   (A.8)

52. After installing the transmission and cranking the engine, a grinding noise is heard. The LEAST-Likely cause of this is:
    A. a starter alignment that needs to be adjusted.
    B. missing engine dowels.
    C. a bent or rubbing inspection cover.
    D. a worn transmission main shaft bearing.   (A.8)

53. When checking flywheel runout, what should be checked first?
    A. The ring gear
    B. The crankshaft end play
    C. The pilot bushing
    D. Free travel in the clutch pedal   (A.9)

54. After inspection of the synchronizer sleeve for third and fourth gear, the teeth on the sleeve are found to be rounded and worn. What is the LEAST-Likely cause of this wear?
    A. Third gear
    B. Fourth gear
    C. Third and fourth gear blocking rings
    D. Third and fourth synchronizer hub   (C.7)

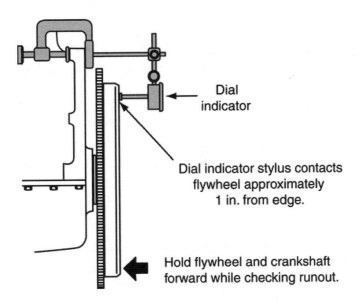

Dial
indicator

Dial indicator stylus contacts
flywheel approximately
1 in. from edge.

Hold flywheel and crankshaft
forward while checking runout.

55. In the figure, Technician A says crankshaft end play can be checked with the tool in the above
location. Technician B says that with the tool in this position you can check flywheel runout.
Who is right?
   A. A only
   B. B only
   C. Both A and B
   D. Neither A nor B   (A.9)

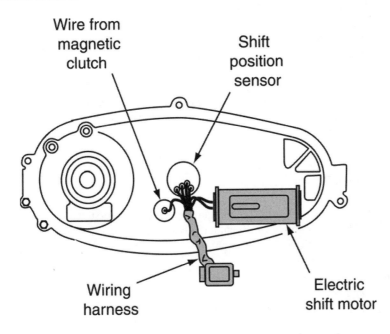

Wire from
magnetic
clutch

Shift
position
sensor

Wiring
harness

Electric
shift motor

56. The best way to check mating surfaces for warpage on a transaxle case is to use a:
   A. straightedge.
   B. dial indicator
   C. micrometer
   D. flat surface   (C.3)

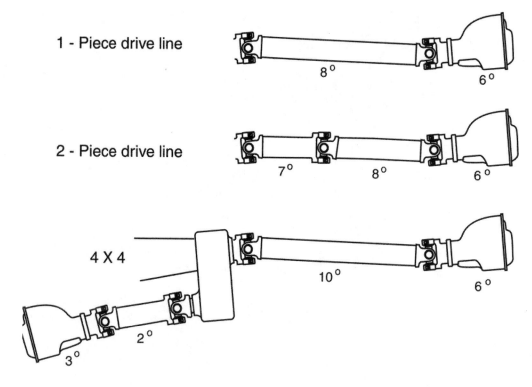

1 - Piece drive line

2 - Piece drive line

4 X 4

57. If a driveshaft has too great an angle between point A and point B in the figure shown, which components are MOST Likely to be damaged?
   A. The spider gears
   B. The transmission mount
   C. The U-joints
   D. The rear tires   (D.6)

58. Technician A says that when checking rear axle runout, rotate the axle slowly. Technician B says to use a dial indicator to measure runout. Who is right?
   A. A only
   B. B only
   C. Both A and B
   D. Neither A nor B   (E.4.4)

59. When installing a new clutch disc, Technician A says that the torsional dampening springs on the clutch should face the flywheel. Technician B says that torsional dampening springs help smooth out engine pulsations. Who is right?
   A. A only
   B. B only
   C. Both A and B
   D. Neither A nor B   (A.5)

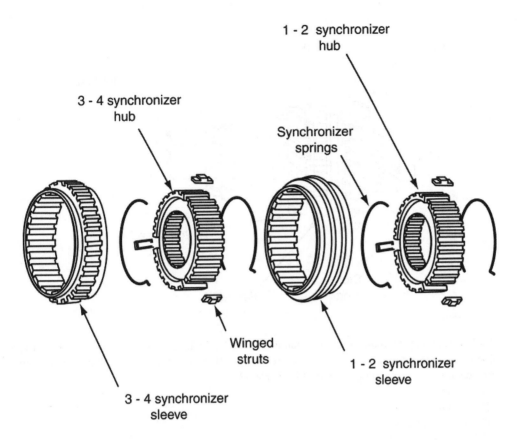

1 - 2 synchronizer hub

3 - 4 synchronizer hub

Synchronizer springs

Winged struts

1 - 2 synchronizer sleeve

3 - 4 synchronizer sleeve

60. A shift fork is connected to the:
    A. forward and reverse gears.
    B. counter shaft.
    C. blocker rings.
    D. synchronizer sleeve.   (C.6)

61. The distance from the rivet heads to the clutch facing surface should be no less than:
    A. 0.005 inch (0.127 mm).
    B. 0.008 inch (0.203 mm).
    C. 0.012 inch (0.305 mm).
    D. 0.025 inch (0.638 mm).   (A.5)

62. When replacing a clutch, the pressure plate is found to have small cracks. Technician A says the pressure plate should be replaced. Technician B says the pressure plate should be resurfaced, then installed. Who is right?
    A. A only
    B. B only
    C. Both A and B
    D. Neither A nor B   (A.5)

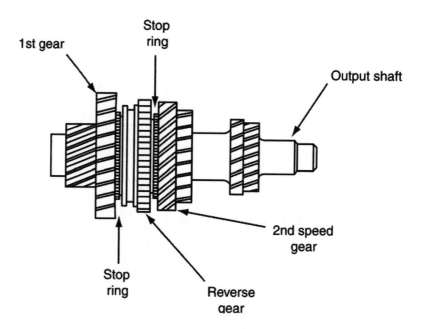

63. While discussing the rebuilding of a manual transaxle, Technician A says that a bent shift fork should be replaced with a new shift fork. Technician B says that the shift fork may be heated and bent back into its original position. Who is right?
    A. A only
    B. B only
    C. Both A and B
    D. Neither A nor B   (C.6)

64. Technician A says that inspecting the ring gear tooth pattern is important because it also checks the carrier bearing preload. Technician B says that inspecting the ring gear tooth pattern is important because it also checks the pinion depth. Who is right?
    A. A only
    B. B only
    C. Both A and B
    D. Neither A nor B   (E.1.8)

65. During disassembly, mark the position of the flywheel in relation to the:
    A. transmission input shaft.
    B. crankshaft.
    C. clutch disc.
    D. bell housing.   (A.7)

66. Technician A says that a worn ring gear on the flywheel can cause clutch wear. Technician B says that a worn ring gear on the flywheel can affect the starter. Who is right?
    A. A only
    B. B only
    C. Both A and B
    D. Neither A nor B   (A.7)

67. Technician A says if side gears or spider gears are damaged, noise would only be heard when the vehicle is turned. Technician B says that side gears are shimmed for proper clearance. Who is right?
   A. A only
   B. B only
   C. Both A and B
   D. Neither A nor B   (C.14)

68. Technician A says that differential noises can be mistaken for engine noises, wheel bearing noises, or tire noises. Technician B says differential noise will occur when the vehicle is moving. Who is right?
   A. A only
   B. B only
   C. Both A and B
   D. Neither A nor B   (C.14)

69. A four-speed transaxle has a clunking noise when driven in first gear and reverse. Technician A says that the gear on the counter shaft could be the problem. Technician B says that the reverse idler gear could be the problem. Who is right?
   A. A only
   B. B only
   C. Both A and B
   D. Neither A nor B   (C.7 and C.10)

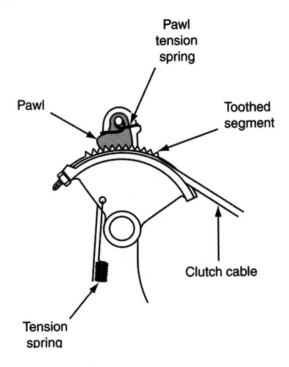

70. A clutch with a self-adjusting cable has:
   A. 1 inch (25.4 mm) of clutch pedal free play.
   B. 2 inches (50.8 mm) of clutch pedal free play.
   C. a constant running release bearing.
   D. an overcenter assist spring.   (A.2)

71. Technician A says that a clutch may slip when it is out of adjustment. Technician B says that a transmission may grind when the clutch is out of adjustment. Who is right?
   A. A only
   B. B only
   C. Both A and B
   D. Neither A nor B   (A.2)

72. On a vehicle with a manual transaxle that jumps out of second gear, all of the following could be the cause **EXCEPT:**
   A. a worn second gear blocking ring.
   B. an excessive main shaft end play.
   C. a shifter linkage out of adjustment.
   D. a worn ring gear.   (C.8)

73. While discussing the rebuilding of a transaxle, Technician A says that snap rings and spacers are usually replaced. Technician B says that snap rings and spacers usually cannot be obtained in a small kit, and must be obtained in a complete overhaul kit. Who is right?
   A. A only
   B. B only
   C. Both A and B
   D. Neither A nor B   (C.8)

74. A growling noise is heard during deceleration. The LEAST-Likely cause of this could be:
   A. a worn pinion bearing.
   B. a worn side gear bearing.
   C. a worn axle bearing.
   D. worn spider gears.   (E.1.1)

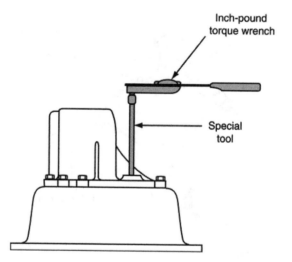

75. When the turning torque in the illustration shown is less than specified:
   A. a thicker shim should be installed behind the side bearing cup in the bell housing side of the case:
   B. a thinner shim should be installed behind both differential side bearing cups.
   C. a thinner shim should be installed behind both differential side bearings.
   D. a thicker shim should be installed behind both differential side gears.   (C.18)

76. Technician A says the driveshaft center support bearing can cause noises in neutral if the vehicle is stopped. Technician B says that some center support bearings need to be lubricated. Who is right?
    A. A only
    B. B only
    C. Both A and B
    D. Neither A nor B   (D.3)

77. While discussing reverse idler gears in a manual transmission, Technician A says that a reverse idler gear may ride on needle bearings. Technician B says that a reverse idler gear may ride on a bronze bushing. Who is right?
    A. A only
    B. B only
    C. Both A and B
    D. Neither A nor B   (C.10)

78. Technician A says that if the countershaft is worn, the bushing inside the counter/cluster gear should be checked. Technician B says that if the shaft is worn, the case and all the reverse gears should be checked. Who is right?
    A. A only
    B. B only
    C. Both A and B
    D. Neither A nor B   (C.10)

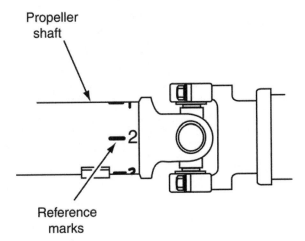

79. The marking shown in the figure is required for:
    A. driveshaft removal and replacement.
    B. universal joint removal and replacement.
    C. driveshaft runout measurement.
    D. driveshaft balance testing.   (D.4)

80. Technician A says that the speedometer drive gear does not have to be replaced when a new speedometer cable core is replaced. Technician B says that by changing the drive gear, the speedometer readings may change. Who is right?
    A. A only
    B. B only
    C. Both A and B
    D. Neither A nor B   (C.12)

81. When removing any transaxle from a vehicle, you will need to:
    A. install an engine support.
    B. drain the engine oil.
    C. disconnect the positive battery cable.
    D. remove the engine.   (C.4)

82. When removing the transaxle differential from the vehicle, the LEAST-Likely component to be removed would be the:
    A. axle shaft.
    B. transaxle.
    C. engine.
    D. lower control arms.   (C.4)

83. The speedometer drive gear is all of the following **EXCEPT:**
    A. mounted on the output shaft.
    B. made of a plastic nylon material.
    C. machined into the output shaft.
    D. a gear with helical teeth.   (C.12)

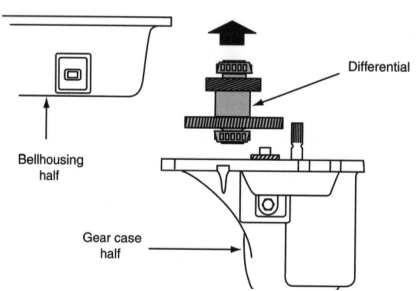

84. In order to remove the differential assembly in a transaxle, you must:
    A. remove the transaxle assembly.
    B. split the transaxle case while it is still in the vehicle.
    C. remove the engine from the vehicle.
    D. disassemble the entire transaxle.   (C.15)

85. The speedometer in a vehicle stopped working, but the sensor circuitry and the speedometer head are OK. Which of the following problems would be the LEAST-Likely cause?
    A. The drive and driven gears are stripped.
    B. The drive gear slips on the end of the sensor.
    C. The drive gear is not positioned correctly on the output shaft.
    D. The driven gear teeth are stripped.   (C.12)

86. The pinion gear rotating torque is too high. To fix this problem:
    A. the bearings will have to be replaced.
    B. the pinion gear will have to be replaced.
    C. the pinion nut can be slightly loosened.
    D. the collapsible spacer must be replaced and the turning torque reset.   (E.1.4 and E.1.6)

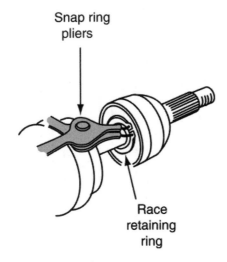

87. Technician A says that a clunking noise while turning could be caused by a bad constant velocity (CV) joint. Technician B says a CV joint could also make noise when the vehicle is traveling straight. Who is right?
    A. A only
    B. B only
    C. Both A and B
    D. Neither A nor B   (D.1)

88. When replacing the ring and pinion gears, all of the following must be replaced **EXCEPT:**
    A. the pinion seal.
    B. the collapsible pinion spacer.
    C. the axle seals.
    D. the spider gears.   (E.1.4)

89. Differential fluid is leaking from an axle housing. Which of the following would be the LEAST-Likely cause?
    A. The axle housing vent is plugged.
    B. The differential is overfilled.
    C. The axle shaft bearings are worn.
    D. The wrong fluid was used and is too thin.   (E.1.1)

90. A new universal joint has been installed in a vehicle. Technician A says the universal joint should be greased until grease comes out of the four caps. Technician B says that the universal joint is prelubricated, and that driveshaft balance affects the amount of grease to install. Who is right?
    A. A only
    B. B only
    C. Both A and B
    D. Neither A nor B   (D.2)

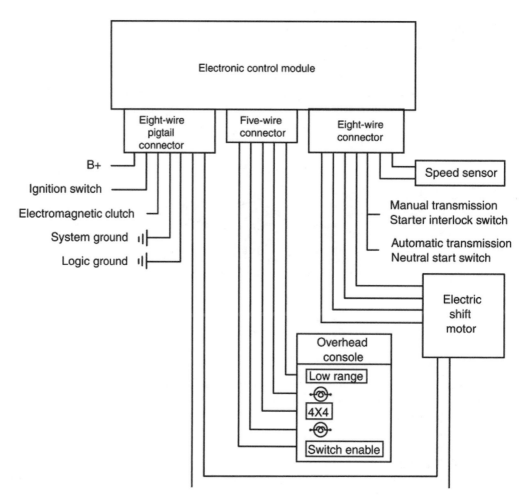

91. Technician A says that all linkages on an electronic-shift transfer case are internal. Technician B says that an open in the shift switch circuit will prevent the transfer case from engaging. Who is right?
    A. A only
    B. B only
    C. Both A and B
    D. Neither A nor B   (F.12)

92. Technician A says an accurate differential case/carrier runout measurement may be performed with scored side bearings. Technician B says the ring gear runout should be measured before the case runout. Who is right?
    A. A only
    B. B only
    C. Both A and B
    D. Neither A nor B   (E.1.3 and E.1.7)

93. To check pinion flange runout:
    A. the driveshaft does not need to be removed.
    B. a straightedge and feeler gauge would be used.
    C. a dial indicator would be used.
    D. the pinion flange will need to be removed.   (E.1.2)

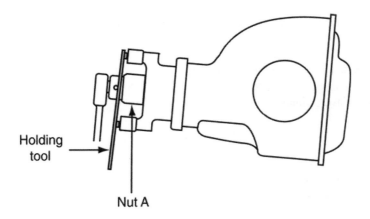

Holding tool →

Nut A

94. In the figure shown, if nut A is loose, which of the following is LEAST Likely to happen?
    A. A clunking noise when vehicle is put in gear
    B. Excess pinion bearing wear
    C. A bad tooth pattern on the ring and pinion gears
    D. Wear on the spider gears   (E.1.2)

95. While discussing differential spider gears, Technician A says the spider gears that ride on the pinion shaft and the bore that runs through the gear should be smooth and shiny. Technician B says that the spider gears ride on the pinion shaft, but the bores in the gears have needle bearings. Who is right?
    A. A only
    B. B only
    C. Both A and B
    D. Neither A nor B   (C.16)

96. All of the following statements about ring gears are true **EXCEPT:**
    A. hunting-type ring and pinion gear sets must be timed.
    B. loose ring gear bolts may cause a gear chuckle or knocking noise while driving the vehicle.
    C. damaged ring gear and pinion gear teeth may cause a ticking noise while driving the vehicle.
    D. the grooved, painted tooth on the pinion gear must be meshed with the painted, notched ring gear teeth on some ring gearsets.   (E.1.3)

97. Technician A says that if the differential is overfilled, it will cause axle seals to leak. Technician B says that worn axle bearings will cause leaks. Who is right?
    A. A only
    B. B only
    C. Both A and B
    D. Neither A nor B   (E.1.1)

98. When draining the fluid from a manual transaxle, a gold-colored material is seen in the fluid. The LEAST-Likely cause is a worn:
    A. blocking ring.
    B. second gear.
    C. thrust washer.
    D. shift fork.   (C.19)

99. A vehicle with a manual transaxle is hard to shift on a cold start only. The cause of this problem could be:
    A. a rusted shift linkage.
    B. the clutch disc sticking to the flywheel.
    C. wrong lubricant in the transaxle.
    D. a bad slave cylinder.   (C.19)

100. Which tool is used to check ring gear runout?
    A. A feeler gauge
    B. A torque wrench
    C. A straightedge
    D. A dial indicator   (E.1.3)

101. To measure the axle shaft end play on the front axles of a 4WD vehicle, all of the following must be removed **EXCEPT:**
    A. the axle locking hubs.
    B. the brake drums.
    C. the brake pads or shoes.
    D. the wheel and tire.   (E.4.4)

102. When measuring pinion bearing preload,:
    A. the axle must be fully assembled.
    B. the axle shafts must be removed.
    C. the differential case/carrier must be removed.
    D. a spring scale with a hook can be used to measure the preload.   (E.1.6)

103. When installing a replacement U-joint that has a grease fitting, the fitting should point toward the:
    A. front of the vehicle.
    B. transmission.
    C. differential.
    D. driveshaft.   (D.1)

104. When replacing an axle boot, which component is LEAST Likely to be inspected visually?
    A. A CV joint
    B. A wheel bearing
    C. An axle shaft
    D. Axle seals   (D.1)

105. Which of the following is usually used to hold differential side bearings in place?
    A. Snap rings
    B. Lockwashers
    C. An interference fit
    D. Axle shaft splines   (C.17 and C.15)

106. When replacing differential side bearings:
    A. they should be packed with grease.
    B. they are put back on with a hammer and punch.
    C. the bearing races should also be replaced.
    D. the differential case/carrier must also be replaced.   (C.17)

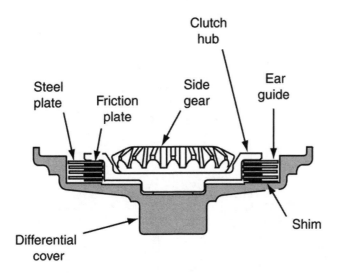

107. Technician A says the left-side differential bearing adjuster nut must be loosened to decrease ring gear backlash. Technician B says pinion preload can be adjusted with selective shims. Who is right?
    A. A only
    B. B only
    C. Both A and B   (E.1.6)
    D. Neither A nor B

108. Two technicians are discussing driveshaft balancing. Technician A says that a strobe light and screw-type hose clamps can be used to balance the driveshaft in the vehicle. Technician B says that driveshaft balancing can only be done with the shaft out of the vehicle. Who is right?
    A. A only
    B. B only
    C. Both A and B
    D. Neither A nor B   (D.4)

109. When balancing a driveshaft, all of the following are true **EXCEPT:**
    A. the driveshaft should be checked for damage before balancing.
    B. screw-type hose clamps can be used for balance weights.
    C. a strobe light is used when checking the balance.
    D. the vehicle suspension should be suspended and not supported.   (D.4)

110. To check the friction plates of a clutch pack:
    A. the friction plates should be measured with a micrometer.
    B. the friction plates should be measured with a feeler gauge.
    C. only a visual inspection is necessary.
    D. the friction plates do not need to be removed from the clutch pack.   (E.3.3)

111. Technician A says 10W-30 oil can be used in a limited slip differential. Technician B says that using the wrong type of oil may cause chatter when the vehicle is cornering. Who is right?
    A. A only
    B. B only
    C. Both A and B
    D. Neither A nor B   (E.3.1)

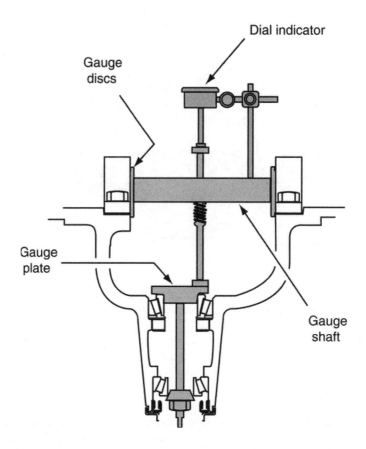

Dial indicator

Gauge discs

Gauge plate

Gauge shaft

112. In the figure shown, after the dial indicator is rotated to the zero position, with the stem on the gauge plate, and then moved off the gauge plate, the dial indicator pointer moves 0.057 inches (1.45 mm) counterclockwise and the pinion gear is marked -4. The proper pinion depth shim is:
   A. 0.039 inch (0.99 mm).
   B. 0.041 inch (1.04 mm).
   C. 0.042 inch (1.07 mm).
   D. 0.043 inch (1.09 mm).   (E.1.5)

113. Technician A says that preload on the pinion gear should be measured with a dial indicator. Technician B says that pinion gear preload can also be measured with a flat feeler gauge. Who is right?
   A. A only
   B. B only
   C. Both A and B
   D. Neither A nor B   (E.1.6)

114. Technician A says that when removing a wheel stud, you can use a hammer to remove it. Technician B says you should never use a torch to remove a wheel stud. Who is right?
   A. A only
   B. B only
   C. Both A and B
   D. Neither A nor B   (E.4.2)

115. All of the following statements about limited slip differentials are true **EXCEPT:**
     A. friction plates are splined to the sidegears.
     B. steel plates are splined to the rear axle shafts.
     C. each clutch set contains a preload spring.
     D. a special lubricant is required.   (E.3.3)

116. A rear-wheel drive vehicle has a vibration that increases in relation to vehicle speed. Technician
     A says the balance pad may have fallen off the driveshaft. Technician B says some of the
     wheels may be out of balance. Who is right?
     A. A only
     B. B only
     C. Both A and B
     D. Neither A nor B   (D.4)

117. When measuring clearance between the blocking ring and the gear, you should use which of
     the following tools?
     A. A micrometer
     B. A dial indictor
     C. A feeler gauge
     D. A ruler   (B.9)

118. All of the following statements about differential case/carrier and ring gear removal and
     replacement are true **EXCEPT:**
     A. the ring gear runout should be measured before removal of the case/carrier and ring gear
         assembly.
     B. the case/carrier side play should be measured before removal of the case/carrier and ring
         gear assembly.
     C. the side bearing caps should be marked in relation to the housing before removal of the
         case/carrier and ring gear assembly.
     D. the side bearing should be clean and dry before installation of the case/carrier and ring
         gear assembly.   (E.1.7)

119. When draining a limited-slip differential, friction material is found in the fluid. What is the
     MOST-Likely cause?
     A. Damaged axle seals
     B. A damaged pinion seal
     C. Disc wear
     D. Damaged gaskets   (E.3.2)

120. When replacing an extension housing seal, which of the following does not need to be inspected?
     A. The extension housing bushing
     B. The front driveshaft yoke
     C. The extension housing gasket
     D. The input bearing   (B.3)

121. Technician A says that when installing a cork gasket, no added sealant is required. Technician B
     says that rubber gaskets should be installed without any added sealant. Who is right?
     A. A only
     B. B only
     C. Both A and B
     D. Neither A nor B   (B.3)

122. When draining the manual transaxle oil, all of the following should be checked **EXCEPT:**
    A. metallic material in the fluid.
    B. the type of the oil and condition.
    C. leaks.
    D. filter condition.   (C.19)

123. While discussing clutch assemblies, Technician A says that when replacing a clutch assembly, the flywheel should only be resurfaced if it has excessive runout, scoring, or any other types of imperfections on the face. Technician B says that even if the flywheel looks OK, the flywheel should be removed and resurfaced every time the clutch assembly is replaced. Who is right?
    A. A only
    B. B only
    C. Both A and B
    D. Neither A nor B   (A.7)

124. Excessive noise coming from the transfer case may be caused by all of the following **EXCEPT:**
    A. low fluid level.
    B. misalignment of the transfer chain.
    C. a worn universal joint.
    D. a damaged output shaft bearing.   (F.1)

125. When discussing an electronically shifted transfer case coupled to an automatic transmission, Technician A says the transfer case will shift into 4L at any speed. Technician B says the transfer case will shift into 4L with the transmission in any gear. Who is right?
    A. A only
    B. B only
    C. Both A and B
    D. Neither A nor B   (F.5)

7 Appendices

Answers to the Test Questions for the Sample Test Section 5

1.	C	23.	A	45.	C	68.	C
2.	C	24.	C	46.	C	69.	C
3.	B	25.	B	47.	B	70.	A
4.	C	26.	B	48.	A	71.	D
5.	B	27.	C	49.	A	72.	B
6.	D	28.	D	50.	C	73.	D
7.	D	29.	A	51.	C	74.	A
8.	C	30.	D	52.	B	75.	A
9.	A	31.	C	53.	C	76.	D
10.	A	32.	B	54.	C	77.	B
11.	D	33.	A	55.	A	78.	B
12.	C	34.	B	56.	D	79.	B
13.	D	35.	C	57.	A	80.	D
14.	C	36.	A	58.	B	81.	B
15.	D	37.	C	59.	D	82.	A
16.	A	38.	C	60.	C	83.	C
17.	A	39.	A	61.	B	84.	A
18.	D	40.	A	62.	C	85.	C
19.	C	41.	D	63.	D	86.	B
20.	A	42.	D	64.	D	87.	C
21.	A	43.	D	65.	B	88.	C
22.	B	44.	D	66.	A	89.	B
				67.	A		

Explanations to the Answers for the Sample Test Section 5

Question #1

Answer A is wrong. The transfer case usually will shift without fluid.

Answer B is wrong. The transfer case will shift into 4WD, but it may vibrate severely.

Answer C is correct. The manual linkage could be seized.

Answer D is wrong. A manual shift transfer case does not have an electronic shift motor.

A range control linkage adjustment is required on some transfer cases. Place the range control lever in the 2WD position, and place the specified spacer between the gate in the console and the lever. Place the outer lever on the transfer case in the 2WD position. Adjust the linkage so it is in the specified location in relation to the outer lever. Regular maintenance is needed on external shift linkage due to a four-wheel drive (4WD) vehicle linkage and transfer case is submerged in water many times. External shift linkages should be lubricated frequently.

Question #2

Answer A is wrong. Most inner tripods are held in place with snap-rings.

Answer B is wrong. Special pliers are used to crimp the outer boot clamps.

Answer C is correct. All of the supplied grease should be used.

Answer D is wrong. Worn CV joints can make a clicking noise when cornering because of lack of lubricant.

Many AWD and 4WD front axle joints have an inner tripod joint. A snap-ring holds the inner spider assembly onto the shaft in this type of joint. Remove the outer boot clamps and slide the boot toward the center of the axle. Spread the retaining ring with a pair of snap-ring pliers, and then slide the outer CV joint from the axle shaft.

Inspect the inner and outer joint components for looseness, wear, scoring, and damage. Check the axle boots for cracks, oil soaking, and deterioration. Replace the boot and axle joints as required. Always install all of the grease supplied with the replacement joints in the joint. Always install the boot clamps using the manufacturer's recommended procedure. Some outer boot clamps must be installed with a special swaging tool.

Question #3

Answer A is wrong. The shift control motor is operated by the GEM.

Answer B is correct. The shift control motor is operated by the GEM.

Answer C is wrong. The shift control motor is operated by the GEM.

Answer D is wrong. The shift control motor is operated by the GEM.

The generic electronic module (GEM) analyzes input information to determine if the proper design conditions are present to provide the driver-requested shift. These inputs include the transfer case motor sense plate, transmission range (TR) sensor or neutral safety switch, and vehicle speed sensor (VSS). The transfer motor sense plate sends a signal to the GEM in relation to the transfer case shift motor position. The TR sensor informs the GEM if the transmission is in neutral or some other gear. The VSS signal informs the GEM regarding vehicle speed. When the inputs indicate to the GEM that all design conditions are met, the GEM operates the transfer case shift motor to provide the driver-requested shift. The transfer case shift motor drives a rotary shift cam, which moves the mode shift fork or range shift fork to the selected position. After a transfer case shift is completed, the shift position sensor sends a signal to the GEM regarding shift motor position.

Question #4
Answer A is wrong. In the A4WD mode, if the front driveshaft begins to turn faster than the rear driveshaft, the computer increases the clutch coil duty-cycle.
Answer B is wrong. In the A4WD mode, if the front driveshaft begins to turn faster than the rear driveshaft, the computer increases the clutch coil duty-cycle.
Answer C is correct. In the A4WD mode, if the front driveshaft begins to turn faster than the rear driveshaft, the computer increases the clutch coil duty-cycle.
Answer D is wrong. In the A4WD mode, if the front driveshaft begins to turn faster than the rear driveshaft, the computer increases the clutch coil duty-cycle.
Some electronically shifted transfer cases have an automatic four-wheel drive (A4WD) function. In these systems the driver may select (A4WD) 4WD High or 4WD Low. Hall-effect sensors monitor both input-shaft and output-shaft speed. The GEM operates the transfer case clutch with a variable duty cycle. During normal cruise driving with the selector switch in A4WD, the GEM operates the transfer case clutch with a minimum duty cycle. This action allows for a slight difference in speed between the front and rear differentials. If the Hall-effect input sensors indicate a specific difference between input- and output-shaft speed, the GEM increases the clutch duty cycle to eliminate the wheel slip that is causing the difference in speed.

Question #5
Answer A is wrong. A hydraulic clutch system does not require free play.
Answer B is correct. Air in the system will prevent the clutch from disengaging properly.
Answer C is wrong. If the clutch facings were worn, it would have no effect on clutch disengagement.
Answer D is wrong. A scored pressure plate would not cause clutch disc disengagement problems.
Some vehicles have linkages connected from the clutch pedal to the release fork. In other clutch systems, the clutch pedal is connected to the release fork by a cable. Another popular method designed to operate the clutch is a hydraulically controlled system. Many clutch linkages or cables have an adjustment to set the clutch pedal free play. Clutch pedal free play is the amount of pedal movement before the release bearing contacts the pressure plate release fingers or diaphragm. Hydraulic clutches must be checked for air in the hydraulic system. This may prevent the clutch from fully disengaging and cause gear clash. Many hydraulic clutch systems must be reverse bled to remove all air from the system and provide a full pedal.

Question #6
Answer A is wrong. Technician A is wrong. On a rear-wheel drive vehicle, the axle flange is part of the axle shaft; on a front-wheel drive vehicle it is unbolted.
Answer B is wrong. Technician B is wrong. Heating the wheel studs can damage the axle flange or the related components.
Answer C is wrong. Neither Technician is correct. If the threads on the axle studs are damaged, run a die over the threads. When these studs are damaged or bent, they should be replaced. Remove the axle and use a hydraulic press to remove and replace the studs. Never apply heat to wheel studs because it will change the metal characteristics and could lead to a failure of the studs.
Answer D is correct. Neither Technician is correct.

Question #7
Answer A is wrong. Neither Technician is correct.
Answer B is wrong. Neither Technician is correct.
Answer C is wrong. Neither Technician is correct. Four-wheel-drive vehicles may have manual or automatic locking hubs. After the wheel bearings have been adjusted and the locking screws or nuts tightened to the specified torque, place some multi-purpose grease on the hub inner spines. Do not pack the hub with grease. Slide the locking hub assembly into the wheel hub until it seats. Install the lock ring in the wheel hub groove. Install the lock washer and axle shaft stop on the axle bolt; then install the bolt and tighten it to the specified torque. Place a small amount of lubricant on the cap seal, and install the cap over the body and into the wheel hub. Install the attaching bolts and tighten them to the specified torque. Rotate the locking hub control from stop to stop to make sure it operates freely. Set both controls to the same auto or lock position.
Answer D is correct. Neither Technician is correct. Automatic locking hubs are lubricated with oil, not packed with grease.

Question #8
Answer A is wrong. First gear is at the rear of the transmission, away from the input shaft.
Answer B is wrong. Second gear is the second closest gear to the rear of the transmission.
Answer C is correct. Fourth gear is the closest to the input shaft.
Answer D is wrong. Reverse is at the rear of the transmission.
Some vehicle manufacturers provide service procedures and specifications for measuring the end play on various transmission gears. In many end play adjustments, a dial indicator is mounted on the transmission case, and the dial indicator stem is positioned on the gear. Push the gear back and forth and observe the end play reading on the dial indicator. Excessive gear end play is usually caused by worn thrust washers.

Question #9
Answer A is correct. Only Technician A is correct. The gasket or sealer will seal the mating surfaces if the burrs have been filed off.
Answer B is wrong. The gasket or sealer will seal the mating surfaces if the burrs have been filed off.
Answer C is wrong. Only Technician A is correct.
Answer D is wrong. Only Technician A is correct. Inspect the extension housing for cracks, and check mating surfaces of this housing and the transmission case for metal burrs and gouges. Remove any metal burrs with a fine-toothed file. Replace the extension housing seal and check the extension housing bushing for wear or damage.

Question #10
Answer A is correct. If the shaft has damage in the seal area, it should be replaced.
Answer B is wrong. It is always a good practice to replace the axle seals while the axle shafts are removed.
Answer C is wrong. Always stand the axle shaft on its end so that the splines do not contact the floor, which could cause damage.
Answer D is wrong. Some require the use of a slide hammer to remove them from the differential.
Visually inspect axle shafts for any wear, pitting, or damage that may cause seal or bearing failure. Then perform a more detailed check after the visual inspection. Place a dial indicator stem against the axle flange to measure flange runout and end play. Maintain a slight inward pressure on the axle flange and turn the dial indicator to zero. Observe the flange runout while rotating the axle shaft one revolution. If the runout is more than specified, replace the axle shaft.
Move the axle inward and outward and observe the dial indicator reading to measure the axle shaft end play. If the end play exceeds specifications, inspect the bearing, retainer plate, or C-lock, and C-lock groove in the axle.

Question #11
Answer A is wrong. A hydraulic clutch has its own master cylinder.
Answer B is wrong. The slave cylinder is connected to the clutch master cylinder by hydraulic tubing and hose.
Answer C is wrong. Neither Technician is correct. The hydraulic system for a hydraulic clutch is totally separate from the brake system. The clutch master cylinder is mounted to the firewall in the engine compartment. A line runs from the clutch master cylinder to the slave cylinder on the bell housing, or inside the bell housing on some models. A rod connected to the clutch pedal goes through the firewall into the clutch master cylinder. When the rod is pushed into the clutch master cylinder, it forces fluid through the line, which actuates the slave cylinder to release the clutch. Air in the system will prevent the clutch from disengaging properly when the clutch pedal is fully depressed. Air is compressible; it will compress in the hydraulic system before the spring pressure in the pressure plate releases the clutch disc from the flywheel. The conditions of less free play, worn facings, or a scored pressure plate will cause clutch problems, but will not affect the release of the clutch.
Answer D is correct. Neither Technician is correct.

Question #12

Answer A is wrong. Excessive crankshaft end play does not cause clutch chatter.

Answer B is wrong. Loose main engine bearings do not cause clutch chatter.

Answer C is correct. A scored pressure plate will not apply an even force to the clutch disc.

Answer D is wrong. Improper pressure plate-to-flywheel position would not necessarily cause clutch chatter.

A straightedge and a feeler gauge should be used to measure pressure plate warpage. A warped, scored flywheel or pressure plate will cause clutch pulsations and chatter. If the flywheel does not have locating dowels, always punch mark the pressure plate and flywheel to ensure the pressure plate is reinstalled in the original position. The will reduce the chances of a vibration being induced into the clutch system.

Question #13

Answer A is wrong. The input shaft bearing should be checked.

Answer B is wrong. The input shaft bearing retainer and seal should be checked.

Answer C is wrong. The crankshaft should be checked.

When the clutch is engaged, the transmission input shaft rotates at the same speed as the engine flywheel and pilot bearing at all times. When the clutch is disengaged, the flywheel and pilot bearing rotate on the end of the transmission input shaft and turn faster than the shaft.

If a bushing-type pilot bearing is lubricated with bearing grease, friction actually will increase between the bushing and the transmission input shaft. Lubricate a bushing-type pilot bearing with motor oil. Lubricate a roller-type pilot bearing with wheel bearing grease.

Answer D is correct. The output shaft bushing does not contact the pilot bushing.

Question #14

Answer A is wrong. Technician B is also correct. The clutch and pressure plate will be sitting closer to the engine, and this may be too far for the slave cylinder to operate.

Answer B is wrong. Technician A is also correct. Material is removed from the flywheel when resurfaced, and this would allow the clutch disc to sit closer to the flywheel bolts.

Answer C is correct. Both Technicians are correct.

Answer D is wrong. Both Technicians are correct. If the flywheel is scored, burned, worn, or has heat cracks, it should be resurfaced or replaced. Heat cracks almost always require flywheel replacement. During resurfacing of a flywheel, 0.0010 -0.040 inch (.025mm- 1mm) of material can be safely removed from the flywheel surface. Do not remove excessive material from the flywheel surface.

Question #15

Answer A is wrong. Misalignment would not affect the free play but it would cause premature throwout bearing wear.

Answer B is wrong. A growling noise when the clutch pedal is depressed would indicate that the throwout bearing is worn.

Answer C is wrong. After the clutch is fully engaged, there would not be any vibration.

The mating surfaces of the engine block and bell housing should be inspected for metal burrs and accumulation of foreign material. Clean the mating surfaces with an approved cleaning solution. Remove any metal burrs with a fine-toothed file. Follow the same procedure to clean and inspect the mating surfaces of the bell housing and transmission, including the bell housing bore.

When bell housing face squareness is measured, a dial indicator is attached to a special tool mounted in the center of the clutch disc splined opening, with the dial indicator stem contacting the bell housing face. The tool is rotated to obtain the measurement. If the bell housing runout is excessive, shims may be installed between the bell housing and engine block to correct this misalignment.

When performing the bell housing bore runout measurement, a dial indicator is attached to a special tool mounted in the center of the clutch disc splined opening with the dial indicator stem contacting the bell housing bore. The tool is rotated to obtain the measurement. Some engines have offset dowel pins in the block surface that mates with the bell housing surface. These offset dowel pins may be rotated to correct the bell housing runout. If the offset dowel adjustment is not enough to correct the bell housing bore runout, replace the bell housing.

Answer D is correct. Misalignment would allow one part of the clutch to grab sooner than the rest of the clutch disc.

Question #16
Answer A is correct. This is the only choice in which the input shaft and pilot bearing are turning at different speeds to cause a rattle.
Answer B is wrong. While decelerating in high gear, the load on the pilot bearing is not at its greatest.
Answer C is wrong. While accelerating in low gear, the load on the pilot bearing is not at its greatest.
Answer D is wrong. There is almost no load on the pilot bearing while the engine is in neutral and the clutch engaged.
The pilot bearing/bushing is mounted in the rear of the crankshaft at the center of the flywheel. This bearing/bushing supports the front of the input shaft, and the clutch disc is splined to the input shaft. The pilot bearing may be a ball bearing, needle bearing, or a bushing. The proper puller should be used to remove the bearing/bushing. The proper driving tool must be used to install the pilot bearing/bushing.

Question #17
Answer A is correct. Only Technician A is correct. In some situations, motor mounts can cause the engine/transmission angle to change causing driveshaft working angle problems.
Answer B is wrong. A driveshaft vibration will not have a constant speed because the driveshaft speed is always changing with incorrect U-joint angles.
Answer C is wrong. Only Technician A is correct.
Answer D is wrong. Only Technician A is correct. Raise the vehicle on a lift so the rear wheels are free to rotate. Clean out the surface of the bearing caps in the front and rear U-joints. Be sure one of the rear U-joint bearing caps is facing straight down. Install the magnetic end of the inclinometer adjusting knob until the weighted cord is centered on the scale. Remove the inclinometer and turn the driveshaft 90 degrees. Install the inclinometer on the U-joint bearing cap facing downward, and record the degree reading where the weighted cord appears on the scale. The driveshaft angle is the difference between the zero reading and the reading when the shaft is rotated 90 degrees.
Repeat the procedure on the front U-joint to obtain the front driveshaft angle. If the driveshaft angles are not within specifications, inspect the engine and transmission mounts for proper position, wear looseness, breaks, and deterioration. Inspect the rear suspension mounting bushings and arms for wear, looseness, or a bent condition. Shims may be used at the rear suspension and/or transmission mounts to allow the angles to be adjusted a slight amount on many vehicles.

Question #18
Answer A is wrong. Worn U-joints can cause driveline vibrations.
Answer B is wrong. Worn front-drive axle joints will cause vibrations when turning corners.
Answer C is wrong. Incorrect driveshaft angles will cause vibrations and premature wear of the U-joints.
Worn U-joints may cause a squeaking or clunking noise, and a vibration while driving straight ahead.
Worn outer front drive axle joints on a 4WD vehicle may cause a vibration while cornering.
Answer D is correct. Slip joint spline wear is uncommon and unlikely to cause noticeable vibration.

Question #19
Answer A is wrong. Both Technicians are correct because with the engine at idle and the transmission in neutral with the clutch released, the countershaft will spin. This will create a growling from the countershaft and bearings that are pitted and scored.
Answer B is wrong. Both Technicians are correct because the countershaft spins in all gears and will create a growling from the countershaft and bearings that are pitted and scored.
Answer C is correct. Both Technicians are correct.
Answer D is wrong. Both Technicians are correct. Inspect all the gear teeth on the countershaft/cluster main shaft gears for chips, cracks, pitting, and wear.
A normal gear tooth wear pattern should appear as a polished finish with little wear on the gear face.
Inspect the bearing surfaces in the countershaft/cluster gear bore for roughness, pitting, scoring, or overheating.
Inspect all bearing surfaces on the countershaft/cluster gear shaft for wear, roughness, pitting, and scoring in the bearing contact area. All countershaft/cluster gear needle or ball bearings should be inspected for roughness and looseness. Replace bearings if these conditions are present. Inspect all countershaft/cluster gear thrust washers and retainers for wear.

Question #20

Answer A is correct. A worn extension housing bushing allows the driveshaft yoke to have excessive runout; this will distort the extension housing seal and will allow it to leak.

Answer B is wrong. The countershaft is not in direct contact with the extension housing bushing or seal, and it will have no effect on the components.

Answer C is wrong. Technician B is wrong. The countershaft is not in direct contact with the extension housing bushing or seal, and it will have no effect on the components.

Answer D is wrong. Technician B is wrong. The countershaft is not in direct contact with the extension housing bushing or seal, and it will have no effect on the components. However, Technician A is correct, a worn extension housing bushing allows the driveshaft yoke to have excessive runout; this will distort the extension housing seal and will allow it to leak. The extension housing seal rides on the driveshaft yoke and the yoke is supported by the extension housing bushing. A worn extension housing bushing allows the driveshaft yoke to have excessive runout; this will distort the extension housing seal and will allow it to leak. A leaking extension housing seal should always make you suspicious of the extension housing bushing. The extension housing bushing can be easily replaced. If the outside surface of the slip yoke is damaged, scored, or rough, it will also cause the seal to have repetitive failures.

Question #21

Answer A is correct. Worn, chipped teeth on the third gear on the input shaft would make this type of noise because the noise is only heard in third gear.

Answer B is wrong. After the transmission was shifted into third gear, there would be no noise.

Answer C is wrong. Worn dogteeth on the blocker ring would cause shifting problems only.

Answer D is wrong. Only shifting problems would occur.

Inspect all the gear teeth on the main shaft gears for chips, pitting, and wear. A normal gear tooth wear pattern appears as a polished finish with little wear on the gear face. Inspect the bearing surfaces in the gear for roughness, pitting, and scoring.

Question #22

Answer A is wrong. A bent shift rail usually will not cause the transmission to jump out of gear.

Answer B is correct. A bent shift rail usually causes hard shifting.

Answer C is wrong. A bent shift rail will not usually cause gear clash during shifting.

Answer D is wrong. A bent shift rail will not cause gear noise.

Shift rails should be inspected to be sure they are not bent, broken, or worn. Inspect all the holes and notches in the shift rails for wear and damage. When each shift rail is installed in the appropriate bore in the case of the shift cover, check for excessive movement between the rail and the bore. Inspect all detent springs to be sure they are not worn, bent, or weak. Check the interlock plates for flatness with a straightedge.

If the interlock plates are worn on the surface that contacts the shift rails, replace the plates.

Question #23

Answer A is correct. Removing shims from the right side and adding an equal number to the left moves the ring gear inward to decrease backlash without changing preload.

Answer B is wrong. Adding shims equally to both sides will change preload without affecting backlash.

Answer C is wrong. Removing shims from the left and adding an equal number to the right side moves the ring gear outward to increase backlash.

Answer D is wrong. Removing shims equally from both sides will change preload without affecting backlash.

When shims are positioned behind the side bearings to adjust backlash and preload, the differential case/carrier is pried to one side and the movement is recorded with a dial indicator or feeler gauge. Service spacers, shims and feeler gauges are installed on each side of the side bearing to obtain zero side play and zero backlash. Then calculate the proper shim thickness to provide the specified backlash and sidebearing preload. On some differentials, the proper shims have to be driven into place behind the side bearings with a special tool and a soft hammer.

Question #24
Answer A is wrong. If a pilot bearing is damaged, it will make more noise with the clutch disengaged.
Answer B is wrong. If the input shaft pilot bearing contact surface is damaged, it will make more noise with the clutch disengaged.
Answer C is correct. The input is not turning with the clutch depressed, but is turning with the clutch engaged.
Answer D is wrong. A main shaft bearing failure would make noise no matter what position the clutch was in. The tip of the input shaft that rides on the pilot bearing/bushing must be smooth. Some transaxles are not designed with a pilot bearing/bushing and their input shafts do not have a machined surface on the tip.
The clutch disc rides on the splines of the input shaft. Worn input shaft splines may cause the clutch disc to stick on these splines, resulting in improper clutch release. Be sure the input shaft teeth that mesh with the synchronizer collar are not worn or chipped. Remove the roller bearings from the inner end of the input shaft. These bearings and the bearing contact area in the input shaft must be inspected for wear, pitting, and roughness. Check the input shaft bearing for smooth rotation and looseness.

Question #25
Answer A is wrong. The dogteeth tips should be sharp and pointed.
Answer B is correct. The threads in the cone area of the blocking rings should be sharp and not dull.
Answer C is wrong. There must be specified clearance between the blocker rings and the gear teeth.
Answer D is wrong. The synchronizer sleeve must slide freely on the hub.
Be sure the synchronizer sleeve moves freely on the hub. Inspect the insert springs to be sure they are not bent, distorted, or broken. Inspect the blocking rings for cracks, breaks, and flatness. The dogteeth on the blocking rings must be pointed with smooth surfaces. Rounded, chipped dogteeth will cause gear clash during shifting. Threads in the cone area of the blocking rings must be sharp and not dulled. Some manufacturers utilize a blocking ring lined with friction material; this material must be intact and not worn.
With the blocking ring positioned on the matching gear, measure the distance between the blocking ring and the matching gear's dogteeth with a feeler gauge. Too little clearance will cause gear clash while shifting.

Question #26
Answer A is wrong. If the clearance on the fourth-speed gear is less than specified, it would only affect the transition into fourth gear. It could not cause noise when fourth gear is engaged.
Answer B is correct. Only Technician B is correct. Fourth-gear clearance that is less than specifications can cause hard shifting.
Answer C is wrong. Only Technician B is correct.
Answer D is wrong. Technician B is correct. The blocking ring dogteeth tips should be pointed with smooth surfaces.
Clearance between the blocking ring and the matching gear dogteeth is important for proper shifting. The synchronizer sleeve must slide freely on the synchronizer hub. The threads on the blocking ring in the cone area must be sharp to get a good bite on the gear to stop it from spinning and to make a synchronized non-clashing shift.
If the clearance between the blocking ring and the fourth-speed gear dogteeth is less than specified, the blocking ring is worn, which results in hard shifting. This problem would not result in noise while driving in fourth gear.

Question #27
Answer A is wrong. The yoke seal should keep fluid from appearing at the yoke opening.
Answer B is wrong. A transfer case does not have a sight glass for checking fluid levels.
Answer C is correct. Fluid should just start to come back out of the fill hole.
Answer D is wrong. A piece of mechanic's wire inserted into the fill hole is not a practiced level-checking method.
When filling a transmission or differential case/carrier, fluid will come out of the fill hole as it reaches the full level. After draining differential fluid, inspect it for excessive metal particles. Silver- or steel-colored particles are signs of gear or bearing wear. Copper- or bronze-colored particles are signs of limited slip clutch disc wear.

Question #28
Answer A is wrong. Excessive end play in the counter gear could cause transmission case damage.
Answer B is wrong. Excessive end play in the counter gear could cause gear teeth damage.
Answer C is wrong. Excessive end play in the counter gear could cause countershaft damage.
Damaged counter (cluster) gear bearings may cause a growling noise with the engine idling with the transmission in neutral and the clutch pedal released because the counter (cluster) gear is turning with the clutch pedal released in neutral, and in gear. It will also cause a growling noise while driving in any gear.
All counter (cluster) gears should show wear patterns in the center of their teeth. These wear patterns should appear as a polished finish, with little wear on the gear face. Check the gears' teeth carefully for chips, pitting, cracks, or breakage. Also, inspect the bearing surfaces to make sure they are smooth. Any damage to the assembly requires replacement.
Answer D is correct. Transmission end play has no effect on the throwout bearing.

Question #29
Answer A is correct. Only Technician A is correct. A broken shift fork would not usually cause noise.
Answer B is wrong. If the reverse idler gear teeth were damaged, there would be a noise.
Answer C is wrong. Only Technician A is correct.
Answer D is wrong. Only Technician A is correct. The reverse idler gear teeth should be inspected for chips, pits, and cracks. Check the gear bore for roughness and scoring. The reverse idler shaft and bearings must be inspected for roughness, scoring, and pitting. Inspect all shift forks, shift rail bushings, and replace all shift rail seals. Replace any shift fork that shows any sign of off-center wear

Question #30
Answer A is wrong. The output shaft seal is not the LEAST-Likely component that may need to be replaced.
Answer B is wrong. The tail housing gasket is not the LEAST-Likely component that may need to be replaced.
Answer C is wrong. The speedometer O-ring is not the LEAST-Likely component that may need to be replaced. Check the extension housing for cracks and repair or replace it as needed. Check the mating surfaces of the housing for burrs or gouges and file the surface flat. Inspect the speedometer cable or sensor for any leakage from the seal; replace the seal if fluid is leaking. Install a new gasket to the extension housing during installation. Check all threaded holes and repair any damaged bores with a thread repair kit. Check the bushing in the rear of the extension housing for excessive wear or damage. Always replace the rear extension housing seal
Answer D is correct. The counter gear shaft is housed in the transmission case, not the extension housing.

Question #31
Answer A is wrong. The driven gear is located on the vehicle speed sensor.
Answer B is wrong. The driving gear is located on the output shaft.
Answer C is correct. The driving gear is located on the output shaft.
Answer D is wrong. The speedometer driving gear is not located on the axle shaft.
The speedometer driving gear is the gear located on the output shaft of the transmission, and it can be accessed by removing the rear extension housing. The speedometer driven gear is located on the vehicle speed sensor and is located in the rear extension housing. There are no speedometer gears located on the main shaft or the axle shaft.

Question #32
Answer A is wrong. The remote vent tube on a differential does not increase pressure.
Answer B is correct. The remote vent tube is located above a point where moisture could enter.
Answer C is wrong. The remote vent tube on a differential does not keep lubricant from coming out of the differential.
Answer D is wrong. The remote vent tube on a differential is not used to add lubricant to the differential.
The only purpose of a remote vent is to keep moisture out of a differential assembly. This is needed on a four-wheel drive (4WD) vehicle in case the axle or transfer case is submerged in water. A plugged transfer case vent may cause seal leakage.

Question #33

Answer A is correct. ATF is a thinner viscosity oil than hypoid gear oil, and it reduces friction and improves fuel economy.
Answer B is wrong. The thicker the gear oil, the higher the viscosity number.
Answer C is wrong. Only Technician A is correct.
Answer D is wrong. Technician A is correct. Some vehicle manufacturers recommend a mineral oil with an extreme pressure (EP) additive for manual transmissions. The most common gear oil classifications are SAE 75W, 75W-80, 80W-90, 85W-90, 90, or 140. Thicker gear oils have higher classification numbers. Some manual transmissions require engine oil or automatic transmission fluid. Many manual transaxles require 30W engine oil, 90W gear oil, or ATF. Using the wrong fluid can cause shifting and/or wear concerns. Using too thick oil in a transmission that is designed for 30W oil or ATF can cause shifting concerns during cold weather.

Question #34

Answer A is wrong. A worn fourth-gear synchronizer would only affect shifting into fourth gear.
Answer B is correct. A sticking clutch disc would affect all gears.
Answer C is wrong. A worn blocker ring would not cause clash in all gears.
Answer D is wrong. If only the 3−4 shifter fork were worn, then just shifting into third or fourth gear would cause gear clash. The tip of the input shaft that rides on the pilot bearing/bushing must be smooth. Some transaxles are not designed with a pilot bearing/bushing, and their input shafts do not have a machined surface on the tip.
The clutch disc rides on the splines of the input shaft. Worn input shaft splines may cause the clutch disc to stick on these splines, resulting in improper clutch release. Be sure the input shaft teeth that mesh with the synchronizer collar are not worn or chipped. Remove the roller bearings from the inner end of the input shaft. These bearings and the bearing contact area in the input shaft must be inspected for wear, pitting, and roughness. Check the input shaft bearing for smooth rotation and looseness

Question #35

Answer A is wrong. Both Technicians are correct because if the linkage is not adjusted correctly, it may not shift into gear.
Answer B is wrong. Both Technicians are correct because if the linkage is not adjusted correctly, it may be hard to shift out of gear.
Answer C is correct. Both Technicians are correct.
Answer D is wrong. Both Technicians are correct. Most external shift linkages and cables require adjusting, and a similar adjustment procedure is used on some vehicles. Raise the vehicle on a lift and place the shift lever in neutral to begin the shift linkage adjustment. With a lever-type shift linkage install a rod in the adjustment hole in the shifter assembly. Adjust the shift linkage by loosening the rod-retaining locknuts and move the levers until the rod fully enters the alignment holes. Tighten the locknuts and check the shift operation in al gears. Rod-type shift linkages are adjusted with basically the same procedure as the lever-type linkages. When the alignment pin is in place, adjust the shift rod so the pin slides freely in and out of the alignment hole.

Question #36

Answer A is correct. Only Technician A is correct. A plugged vent can cause enough internal pressure to force fluid past a seal.
Answer B is wrong. An outer drive axle joint does not come into contact with the drive seal.
Answer C is wrong. Only Technician A is correct.
Answer D is wrong. Technician A is correct. Transaxle leaks may occur at the drive axle seals, vent, sealing surfaces between the case sections, or drain plug. During a transaxle overhaul, all seals and gaskets should be replaced. Seal bores should be inspected for metal burrs and scratches. Inspect all seal lip contact areas for roughness or scoring. Prior to installation, coat the outside diameter of each seal case with an appropriate sealer. Always use the proper driver to install each seal. Coat the seal lips and lip contact area with the manufacturer's specified transaxle lubricant.
A plugged vent may result in pressure buildup in the transaxle, also resulting in leaks from the seals.

Question #37
Answer A is wrong. Both Technicians are correct because if the front suspension is not aligned, the drive axles will not be properly aligned.
Answer B is wrong. Both Technicians are correct because the lower control arms are connected to the cradle; if the cradle is not properly aligned, the front suspension will not be properly aligned.
Answer C is correct. Both Technicians are correct.
Answer D is wrong. Both Technicians are correct. Prior to transaxle removal, the battery ground cable, shift linkages or cables, speedometer cable, vehicle speed sensor, and all electrical connections must be disconnected. Drain the transaxle lubricant. On front wheel drive vehicles, the front drive axle must be removed from the transaxle prior to transaxle removal. Before the transaxle retaining bolts are loosened, an engine support fixture must be installed to support the weight of the engine. On some front-wheel-drive vehicles, the engine cradle, or part of the cradle, must be removed prior to transaxle removal. Use a transmission jack to support the weight of the transaxle during the removal process.

Question #38
Answer A is wrong. The transaxle case must be opened before the input shaft can be removed.
Answer B is wrong. The transaxle case must be opened before the output shaft can be removed.
Answer C is correct. The pilot bearing is not part of the transaxle.
Answer D is wrong. The transaxle case must be opened before the differential bearings can be removed.
While assembling a manual transaxle, it is important to apply gear lube to all of the transaxle parts. Before checking the specifications of the shafts in the transaxle, rotate the shafts to work the gear lube into the bearings. A false measurement may be made if the gear lube is not worked into the bearings.
Before disassembling a transaxle, observe the effort it takes to rotate the input shaft through all forward gears and reverse. Extreme effort in any or all gears may indicate an end play problem or a bent shaft.

Question #39
Answer A is correct. Only Technician A is correct. The power flow in the transfer case goes from the input shaft through the sun gear and planetary carrier to provide gear reduction.
Answer B is wrong. The annulus gear is locked to the case so that it cannot rotate.
Answer C is wrong. Only Technician A is correct. The power flow in the transfer case goes from the input shaft through the sun gear and planetary carrier to provide gear reduction.
Answer D is wrong. Only Technician A only is correct. The power flow in the transfer case goes from the input shaft through the sun gear and planetary carrier to provide gear reduction. Specification measurements should be taken and recorded to aid in the installation of parts that have tolerances.
All parts should be cleaned and lubricated before the assembly of the part. All components should be inspected for wear or damage.
The annulus gear is locked to the case so it cannot rotate. In four-wheel drive (4WD) Low, the transfer case input shaft is driving the sun gear, which, in turn, is driving the planetary carrier.

Question #40
Answer A is correct. Only Technician A is correct. Detent springs help to hold the shift rail in the engaged position.
Answer B is wrong. If the fourth gear teeth were worn, the transaxle would jump out of fourth gear.
Answer C is wrong. Only Technician A is correct.
Answer D is wrong. Only Technician A is correct. Shift rails should be inspected to be sure they are not bent, broken, or worn. Inspect all the holes and notches in the shift rails for wear and damage. When each shift rail is installed in the appropriate bore in the case of the shift cover, check for excessive movement between the rail and the bore. Inspect all detent springs to be sure they are not worn, bent, or weak.
Be sure the synchronizer sleeve moves freely on the hub. Inspect the insert springs to be sure they are not bent, distorted, or broken. Inspect the blocking rings for cracks, breaks, and flatness. The dogteeth on the blocking rings must be pointed with smooth surfaces. Rounded, chipped dogteeth will cause gear clash during shifting. Threads in the cone area of the blocking rings must be sharp and not dulled. Some manufacturers utilize a blocking ring lined with friction material; this material must be intact and not worn.
With the blocking ring positioned on the matching gear, measure the distance between the blocking ring and the matching gear's dogteeth with a feeler gauge. Too little clearance will cause gear clash while shifting.

Question #41
Answer A is wrong. A broken detent spring could cause the transmission to jump out of gear.
Answer B is wrong. A broken detent spring could cause harsh shifting.
Answer C is wrong. A broken detent spring could cause the transmission to lock up between gears.
Inspect all detent springs to be sure they are not worn, bent or weak. A weak detent spring on the 3-4 shift rail may cause the transaxle to jump out of third gear. By not having enough spring pressure, the weak spring could cause this to happen.
Answer D is correct. A faulty bearing or bushing will cause a growling noise.

Question #42
Answer A is wrong. The dogteeth only would cause noise when shifting into second gear.
Answer B is wrong. A vibration in second gear would be an indication of gear problems.
Answer C is wrong. Difficult shifting would happen in second gear only.
Synchronizer inspection and replacement is basically the same for transmissions and transaxles. Synchronizer hubs and sleeves are directional. Always assemble these components in their original positions. Prior to disassembly, always mark the synchronizer sleeve and hub so that these components are assembled in their original locations in relation to each other. Use a feeler gauge to measure the clearance between the dogteeth on all speed gears and the matching blocking ring. Replace any blocking ring that indicates a reading lower than the specifications
Answer D is correct. Worn gear dogteeth will not allow the synchronizer sleeve to grip the gear and will allow the transmission to slip out of gear.

Question #43
Answer A is wrong. Technician A is not correct. If the synchronizer is reversed, it would not function.
Answer B is wrong. Technician B is not correct. Synchronizer sleeves are machined to fit in only one way. If reversed, it would not fit together.
Answer C is wrong. Neither Technician is correct. Synchronizer inspection and replacement is basically the same for transmissions and transaxles. Synchronizer hubs and sleeves are directional. Always assemble these components in their original positions. Prior to disassembly, always mark the synchronizer sleeve and hub so that these components are assembled in their original locations in relation to each other.
Answer D is correct. Neither Technician is wrong.

Question #44
Answer A is wrong. The reverse idler gear meshes with the reverse gear and only rotates in reverse.
Answer B is wrong. The reverse idler gear meshes with the reverse gear and only rotates in reverse.
Answer C is wrong. The reverse idler gear meshes with the reverse gear and only rotates in reverse.
The reverse idler gear teeth should be inspected for chips, pits, and cracks. Check the gear bore for roughness and scoring. The needle bearings/bushings and shafts must be inspected for roughness, scoring, and pitting. Replace gears or bearings if these conditions are present.
Answer D is correct. The reverse idler gear meshes with the reverse gear and only rotates in reverse.

Question #45
Answer A is wrong. Speedometer drive gears are never located on the input shaft.
Answer B is wrong. The transfer gear does not rotate at the same rpm as the drive axle.
Answer C is correct. Speedometer drive gears normally operate at output shaft speed.
Answer D is wrong. The drive axle inner hub is not located in the transaxle.
Some manual transaxles have a combined speedometer drive and speed-distance sensor. The speedometer drive gear should be inspected for worn teeth and looseness. Be sure the adapter is not loose on the speed distance sensor extension. A worn adapter may cause erratic speedometer operation. During a transmission overhaul, replace the speedometer drive O-ring. Inspect the speedometer drive gear for worn or damaged teeth and looseness on the differential case/carrier.

Question #46
Answer A is wrong. Technician B is correct. Fuel economy can be affected by using fluid with too much viscosity.
Answer B is wrong. Technician A is correct. The improper fluid type can cause manual transmission leakage. This is why it is so important that the fluid be of recommended type and viscosity.
Answer C is correct. Both Technicians are correct.
Answer D is wrong. Both Technicians are correct. It is important for technicians to include an inspection of drivetrain fluid level and to be sure that manufacturers recommended types of fluids be used. As a technician, you should check manufacturer's TSBs for changes in fluids or additives that have been recommended since the production date. When fluid level is incorrect, it should be adjusted with the correct fluid type to proper levels.

Question #47
Answer A is wrong. VSS generate a signal that cannot be tested with an ammeter.
Answer B is correct. Only Technician B is correct that you can check the VSS output signal with an oscilloscope. Most vehicle speed sensors generate an AC voltage that increases with speed.
Answer C is wrong. Only Technician B only is correct that you can check the VSS output signal with an oscilloscope. Most vehicle speed sensors generate an AC voltage that increases with speed.
Answer D is wrong. Only Technician B only is correct that you can check the VSS output signal with an oscilloscope. Most vehicle speed sensors generate an AC voltage that increases with speed.
Manual transmissions may have various sensors, switches, and solenoids. These components may include a vehicle speed sensor (VSS), a backup light switch, and a computer-aided gear-select solenoid. The VSS sends a voltage signal to the PCM in relation to vehicle speed. The PCM uses this signal to operate output, such as the cruise control. Some vehicles multiplex sensors and obtain the VSS signal from ABS wheel speed sensors. The ABS/VSS sensor is usually a permanent-magnet signal generator. However, some units utilize a Hall-effect device to generate the voltage signal. While some of these types of sensors may be checked using an ohmmeter for initial resistance, the only accurate method of testing the sensors is by using a DMM on the AC voltage setting or by using an oscilloscope.

Question #48

Answer A is correct. The dial indicator measures crankshaft movement forward and backward in the block.
Answer B is wrong. Crankshaft warpage can only be checked with the crankshaft out of the engine.
Answer C is wrong. Main bearing wear has to be measured after the engine is disassembled.
Answer D is wrong. The bell housing has to be installed on the engine block to measure the engine block alignment.
A dial indicator may be positioned against the clutch facing contact area of the flywheel to check crankshaft end play and flywheel runout. Hold the flywheel forward while rotating the flywheel to check flywheel runout. If runout is excessive, resurface or replace the flywheel. To check the crankshaft end play, force the flywheel forward and rearward and measure the travel. The accessory belts should be removed to allow an accurate reading. If runout is excessive, engine thrust bearing service is required.

Question #49

Answer A is correct. Incorrect preload and backlash cause incorrect gear tooth contact, which usually produces noise.
Answer B is wrong. Using the wrong differential lube will not cause a constant whining noise.
Answer C is wrong. If the side gears were damaged, the noise would occur only when turning.
Answer D is wrong. If the spider gears were damaged, the noise would occur only when turning.
Damaged ring gear teeth would cause a clicking noise while the vehicle is in motion.
This problem would not cause differential chatter. Improper preload on differential components, such as side bearings, may cause differential chatter.
If there were a constant whining noise coming from the differential, the noise could not be coming from the side gears or the spider gears. These gears are only used when the vehicle is turning, so if they were damaged, the noise would be heard only on turns. The wrong differential lube could cause damage to the differential parts, but will not cause a whining noise. If the preload and backlash are not set properly, the gear mesh could be too tight and cause a whining noise.

Question #50

Answer A is wrong. The washer thickness has to be calculated individually for each side.
Answer B is wrong. The side gear end play cannot be measured with the shims behind the gears.
Answer C is correct. Select-fit thrust washers are used to set the proper clearance of side gears.
Answer D is wrong. The purpose is to take up excess clearance, not to provide preload.
Reassemble the differential side gears, pinion gears, pinion gear thrust washers, and shaft. Do not install the side gear thrust washers. Install the roll pin/bolt to retain this shaft in the differential case/carrier.
Rotate the side gears two revolutions in each direction. Install a special tool through one of the axle openings against the side gear, and assemble a dial indicator so the stem rests against the tool. Move the side gear up and down and record the end play on the dial indicator. Rotate the side gear 90 degrees and record another end play reading. Turn the side gear another 90 degrees and record a third end play reading. Use the smallest end play recorded to calculate the required side gear washer thickness. The side gear end play must be 0.001 in. to 0-013 in. If the end play recorded was 0.050 in., install a 0.042 in. washer, which provides an end play of 0.008 in.

Question #51

Answer A is wrong. Technician B is also correct. Some, not all, threads in a transaxle can be repaired. See the appropriate service manual.
Answer B is wrong. Technician A is also correct. A helicoil would be the correct way to repair damaged threads in a transaxle.
Answer C is correct. Both Technicians are correct.
Answer D is wrong. Both Technicians are correct. Transaxle case replacement is often required if the case is cracked. Some vehicle manufacturers recommend that the case be repaired, depending on how extensive the damage is. For some transaxle cracks, the transaxle case may be repaired with an epoxy-based sealer. Loctite® is not recommended for repairing any transaxle case. If a threaded area of an aluminum housing is damaged, service kits can be used to insert new threads in the bore. Some threads should never be repaired; check the service manual to identify which ones can be repaired.

Question #52
Answer A is wrong. This choice has nothing to do with the measurement shown.
Answer B is correct. Differential end play measurement requires medium upward pressure and the use of a dial indicator.
Answer C is wrong. The differential shim should not be the same as the end play measurement.
Answer D is wrong. The transaxle bolts should be tightened to the specified torque.
On some transaxles a special puller is used to remove the bearing from the differential case/carrier. These bearings must be installed on the differential case/carrier with a special driving tool. The differential bearing cups are removed from the case/carrier with special tool. A preload shim is located behind one of these bearing cups.
First remove the differential bearing cup and shim from the clutch bell housing half of the transaxle case. Then use the proper driving tool to install a new bearing cup without the shim. Install a new bearing cup in the gear case half of the case with the proper driver. Lubricate the differential bearings with the manufacturer's specified transaxle lubricant. Install the differential assembly and reassemble the case halves. Tighten the case bolts to the specified torque.
Install the special tool into the differential side gears and install a T-handle on top of this tool. Apply downward pressure to the T-handle and rotate this handle several times in both directions to seat the side bearings. Zero the dial indicator installed against the upper side of the differential case/carrier. Apply a medium load to the differential in the upward direction through the opposite drive axle opening. Rotate the differential back and forth with the T-handle while maintaining this upward pressure. Record the end play indicated on the dial indicator.
The proper preload shim thickness is equal to the end play plus 0.007 in. (0.18 mm) to provide sidebearing preload. After the case is assembled, install the special turning tool in the differential. Install an inch-pound torque wrench on this tool and rotate the differential back and forth while observing the turning torque on the torque wrench. If the turning torque is more than specified, reduce shim thickness by 0.002 in. (0.5 mm). When the turning torque is less than specified, increase the shim thickness by 0.002 in. (0.5 mm).

Question #53
Answer A is wrong. Technician B is also correct. A broken oil feeder will not get lubricant to the output shaft and bearings.
Answer B is wrong. Technician A is also correct. Only the correct lubricant will flow through the system as designed and perform as intended.
Answer C is correct. Both Technicians are correct.
Answer D is wrong. Both Technicians are correct. Transmissions usually are splash-lubed; however, some transmissions utilize a metal tube trough that allows oil to build up and run into bearing lubrication passages and maintains an adequate oil supply to the end bearings. Inspect the oil feed trough for damage or a bent condition. A damaged or bent oil feed trough may reduce lubricant flow to the end bearings and cause premature bearing failure. Some output shaft front bearings have an oil feeder behind the bearing to main the proper supply of lubricant to the bearing shaft supported by the bearing. Some vehicle manufacturers recommend a mineral oil with an extreme pressure (EP) additive for manual transmissions. The most common gear oil classifications are SAE 75W, 75W-80, 80W-90, 85W-90, 90, or 140. Thicker gear oils have higher classification numbers. Some manual transmissions require engine oil or automatic transmission fluid. Many manual transaxles require 30W engine oil, 90W gear oil or ATF.
Using the wrong fluid can cause shifting and/or wear concerns. Using too thick oil in a transmission that is designed for 30W oil or ATF can cause shifting concerns during cold weather.

Question #54

Answer A is wrong. Technician B is also correct. The squeaking noise will increase as the vehicle speed increases.

Answer B is wrong. Technician A is also correct. The joint will not follow its intended arc, causing driveline imbalance.

Answer C is correct. Both Technicians are correct.

Answer D is wrong. Both Technicians are correct. A universal joint has prelube on the inside, but this prelube is not sufficient to lubricate the joint when it has been installed in a vehicle. You should grease a new universal joint when it is installed. When a universal joint is greased, you should not pump so much grease into the joint that it squirts out of the caps. When this happens, it damages the seals around the caps and shortens the life of the joint.

In a double Cardan U-joint, the center yoke should be marked in relation to the ball tube yokes prior to disassembly, so that these components may be assembled in their original locations. The bearing caps in this type of joint should be removed in the proper sequence using the same procedure as followed for a single U-joint. The centering ball and ball seats must be replaced if they are worn or scored

Question #55

Answer A is correct. Only Technician A is correct. Bearings must always be lubricated for this adjustment.

Answer B is wrong. If the pinion nut is over-torqued, the assembly has to be disassembled and a new crush sleeve has to be installed before attempting to obtain the proper turning torque.

Answer C is wrong. Only Technician A is correct.

Answer D is wrong. Only Technician A is correct. Coat the outside diameter of the new seal with gasket sealer, and lubricate the seal lips and bearings with the manufacturer's specified differential lubricant. Use the proper seal driver to install the pinion seal.

Install the differential flange and a new pinion nut. Since the pinion nut torque determines the pinion bearing preload, the torque on this nut is critical. Tighten the pinion nut until the manufacturer's specified turning torque is obtained with an inch-pound torque wrench and socket installed on the pinion nut.

Question #56

Answer A is wrong. A bad front axle shaft would not cause a clicking noise while turning.

Answer B is wrong. A constant-velocity inner joint would be affected when the vehicle's suspension rebounds.

Answer C is wrong. A bad torsional damper would cause a shudder in the vehicle, not a clicking sound.

A constant-velocity (CV) inner joint is not affected when the vehicle is turning. It is only affected when the vehicle suspension is jounced or rebounded while driving. Axle shafts are serviceable; therefore the whole axle shaft does not need to be replaced. An outer CV joint will make a clicking noise when the vehicle is turning. This means the joint ball bearings are bad and the grease is contaminated. The CV joint should be replaced.

Answer D is correct. The ball bearings in the outer CV joint are probably damaged.

Question #57

Answer A is correct. Only Technician A is correct. The change in the position and required movement of the bearings inside an outer drive axle joint during a turn will result in a noise that is not present when driving in a straight line.

Answer B is wrong. A wheel bearing usually makes a grinding or whirring noise that may change during turns.

Answer C is wrong. Only Technician A is correct.

Answer D is wrong. Only Technician A is correct. Most CV joints will produce a clicking noise only heard during turns until they are completely worn out. Then they may also make clicking noises while driving straight ahead. They are usually replaced as a complete assembly. Prior to removing the CV joint boot, mark the inner end of the boot in relation to the drive axle so the boot may be reinstalled in the original position. When reassembling the joint, always install the grease in the joint that is provided in the repair kit.

Question #58
Answer A is wrong. If the synchronizer sleeve does not slide smoothly over the blocker ring and gear teeth, shifting will not be smooth, but it will not be prevented.
Answer B is correct. Only Technician B is correct. The synchronizer hub and sleeve must be marked in relation to each other before disassembly to make it work correctly on reassembly.
Answer C is wrong. Only Technician B is correct.
Answer D is wrong. Only Technician B is correct. Be sure the synchronizer sleeve moves freely on the hub. Inspect the insert springs to be sure they are not bent, distorted, or broken. Inspect the blocking rings for cracks, breaks, and flatness. The dogteeth on the blocking rings must be pointed with smooth surfaces. Rounded, chipped dogteeth will cause gear clash during shifting. Threads in the cone area of the blocking rings must be sharp and not dulled. Some manufacturers utilize a blocking ring lined with friction material; this material must be intact and not worn.
With the blocking ring positioned on the matching gear, measure the distance between the blocking ring and the matching gear's dogteeth with a feeler gauge. Too little clearance will cause gear clash while shifting.

Question #59
Answer A is wrong. Technician A is not correct. Transaxles do not use threaded adjusters to adjust the side-bearing preload.
Answer B is wrong. Technician B is not correct. Shims are installed as needed to adjust the side-bearing preload.
Answer C is wrong. Neither Technician is correct. On some transaxles a special puller is used to remove the bearing from the differential case/carrier. These bearings must be installed on the differential case/carrier with a special driving tool. The differential bearing cups are removed from the case/carrier with special tool. A preload shim is located behind one of these bearing cups.
First remove the differential bearing cup and shim from the clutch bell housing half of the transaxle case. Then use the proper driving tool to install a new bearing cup without the shim. Install a new bearing cup in the gear case half of the case with the proper driver. Lubricate the differential bearings with the manufacturer's specified transaxle lubricant. Install the differential assembly and reassemble the case halves. Tighten the case bolts to the specified torque.
Install the special tool into the differential side gears and install a T-handle on top of this tool. Apply downward pressure to the T-handle and rotate this handle several times in both directions to seat the side bearings. Zero the dial indicator installed against the upper side of the differential case/carrier. Apply a medium load to the differential in the upward direction through the opposite drive axle opening. Rotate the differential back and forth with the T-handle, while maintaining this upward pressure. Record the end play indicated on the dial indicator.
Answer D is correct. Neither Technician is correct.

Question #60
Answer A is wrong. The differential cover may be left on while measuring axle shaft end play.
Answer B is wrong. The vehicle does not have to be in neutral to measure axle shaft end play.
Answer C is correct. A dial indicator will measure axle shaft end play at the axle flange, but the brake drums must be removed.
Answer D is wrong. Axle shaft end play cannot be measured with the axle shaft removed from the axle housing.
Place a dial indicator stem against the axle flange to measure flange runout and end play. Maintain a slight inward pressure on the axle flange and turn the dial indicator to zero. Observe the flange runout while rotating the axle shaft one revolution. If the runout is more than specified, replace the axle shaft.
Move the axle inward and outward and observe the dial indicator reading to measure the axle shaft end play. If the end play exceeds specifications, inspect the bearing, retainer plate, or C-lock, and C-lock groove in the axle.

Question #61
Answer A is wrong. This bearing is usually sealed and does not require maintenance.
Answer B is correct. This bearing is usually sealed and does not require maintenance.
Answer C is wrong. This bearing is only used on special applications.
Answer D is wrong. The support bearing is serviceable with the driveshaft removed.
A worn driveshaft center bearing or an outer rear-axle bearing causes a growling noise that is not influenced by acceleration and deceleration. A center support bearing is usually maintenance-free, and it is usually seated. A center support bearing is not part of the driveshaft, but is part of the driveline. A center support bearing is found mainly on trucks and vans and is used to shorten the length of long driveshafts and to decrease pinion angles.

Question #62
Answer A is wrong. Technician B is also correct. A bearing and race are a matched set. A new bearing should never be used with an old race.
Answer B is wrong. Technician A is also correct. If the bearings are not damaged, they may be reused.
Answer C is correct. Both Technicians are correct.
Answer D is wrong. Both Technicians are correct. Ring gear runout and case side play should be measured before removing the ring gear and case assembly. The side-bearing caps should be marked in relation to the case before removal. The side bearings should be lubricated before installation.
The side bearings must be in good condition before measuring case runout. The ring gear runout should be measured before the case runout.
Side-bearing preload limits the amount of lateral differential case/carrier movement in the axle housing or carrier. When the differential has threaded adjusters on the outside of the side bearings, loosen the right adjuster and tighten the left adjuster to obtain zero backlash. Turn the right adjuster the specified amount to obtain the proper preload. Then rotate each adjuster the same amount in opposite directions to obtain the specified backlash.
When shims are positioned behind the side bearings to adjust backlash and preload, the differential case/carrier is pried to one side and the movement is recorded with a dial indicator or feeler gauge. Service spacers, shims and feeler gauges are installed on each side of the side bearing to obtain zero side play and zero backlash. Then calculate the proper shim thickness to provide the specified backlash and sidebearing preload. On some differentials the proper shims have to be driven into place behind the side bearings with a special tool and a soft hammer.

Question #63
Answer A is wrong. Technician A is wrong because a dial indicator should be placed near the center of the driveshaft.
Answer B is wrong. Technician B is wrong. A bent driveshaft should be replaced, not repaired.
Answer C is wrong. Neither Technician is correct. A dial indicator must be mounted to the vehicle underbody near the center of the driveshaft to measure driveshaft runout. Position the dial indicator stem against the surface of the driveshaft. Be sure the driveshaft surface is clean and undamaged. Rotate the driveshaft one revolution to measure the runout. When the runout exceeds specifications, replace the driveshaft and recheck the runout. If the runout is still excessive, check for a bent U-joint flange or slip yoke.
Answer D is correct. Neither Technician is correct.

Question #64
Answer A is wrong. The driveshaft must be installed in the same position from which it was removed.
Answer B is wrong. The transmission cannot be supported by the input shaft; it could bend the clutch disk or input shaft causing clutch chatter or drag; it could fall and damage the transmission or injure the technician.
Answer C is wrong. The engine support fixture should be installed when the transmission bolts are tight. The driveshaft should always be reinstalled in the same position as it was before it was removed. This will reduce the chance of an imbalanced component causing a vibration concern. Always make a reference mark on the driveshaft and pinion yoke before removing the driveshaft. Some manufacturers now place painted dots to match the driveshaft to the pinion yoke, as well as the driveshaft to the transmission output shaft. The weight of the transmission can damage the transmission input shaft and/or the clutch disc.
Answer D is correct. If the clutch disc is not aligned, the transmission input shaft can be damaged and/or the transmission will not be properly aligned.

Question #65

Answer A is wrong. While driving straight ahead, the side gears both turn at the same speed; they would not be the cause of the noise.

Answer B is correct. Only Technician B is correct. Insufficient backlash can cause ring-and-pinion whine.

Answer C is wrong. Only Technician B is correct.

Answer D is wrong. Only Technician B is correct. The pinion bearing preload is measured with an inch-pound torque wrench and socket installed on the pinion nut. Prior to measuring the pinion turning torque with an inch-pound torque wrench, the pinion shaft assembly should be installed with the bearings lubricated and a new collapsible spacer and the proper pinion depth shim. A new pinion shaft nut should be installed. Tighten the pinion nut gradually and keep measuring the turning torque. When the specified turning torque is obtained, the pinion bearing preload is correct. Never loosen the pinion nut to obtain the proper turning torque. If the pinion nut is over-tightened and the turning torque is excessive, install a new collapsible spacer and repeat the procedure.

Inspection and side-play measurement of the side gear, pinion gears, thrust washer, and case/carrier in a rear-wheel-drive differential is basically the same as the inspection of these components in a transaxle differential.

Question #66

Answer A is correct. Only Technician A is correct. Spider gears rotate during cornering.

Answer B is wrong. Spider gears do not rotate at different speeds when the car is going straight.

Answer C is wrong. Only Technician A is correct.

Answer D is wrong. Technician A is correct. To diagnose differential carrier and pinion bearing noises, a stethoscope is often employed to pinpoint the location of the noise. The side gears do not cause a whining noise while driving straight ahead because they are turning only while cornering.

Question #67

Answer A is correct. Only Technician A is correct. A loose pinion nut reduces pinion preload and lets the pinion clunk on the ring gear.

Answer B is wrong. Only a pinion nut that is over-torqued will make a growling noise when the vehicle is in motion.

Answer C is wrong. Only Technician A is correct.

Answer D is wrong. Only Technician A is correct. Before pressing the new rear pinion bearing onto the pinion shaft, be sure the proper spacer washer is installed behind the pinion bearing. The shim determines the pinion depth. When the differential has been disassembled and overhauled, a new collapsible spacer and pinion nut must be installed. Tighten the pinion nut a small amount at a time until the specified turning torque is obtained with an inch-pound torque wrench and socket installed on the pinion nut. Never loosen the pinion nut.

Question #68

Answer A is wrong. A worn blocker ring would only cause a noise during shifting.

Answer B is wrong. A worn first-gear synchronizer ring would only cause a noise during shifting.

Answer C is correct. A chipped and worn first-speed gear would produce a growl in first gear only.

Answer D is wrong. Worn main shaft bearings would cause noise in all gears. If one bearing is bad, it will tend to make more noise when the gears closest to it are engaged, getting quieter as gears further away are engaged.

Inspect all the gear teeth on the main shaft gears for chips, pitting, and wear. A normal gear tooth wear pattern appears as a polished finish with little wear on the gear face. Inspect the bearing surfaces in the gear for roughness, pitting, and scoring. Replace any gears if these conditions are present.

Question #69

Answer A is wrong. Technician B is also correct. Too much lubricant may leak from the axle seals.

Answer B is wrong. Technician A is also correct. The axle vent will allow air to escape from the axle and not have any pressure buildup.

Answer C is correct. Both Technicians are correct.

Answer D is wrong. Both Technicians are correct. Inspect the front and rear output shaft flanges in the transfer case for evidence of oil leakage in the seal area. Drain the fluid to check for water contamination. If there is any sign of leakage at the driveshaft flange seals, these seals must be replaced. When the flanges and seals are removed, always inspect the seal contact area on the flanges for roughness and scoring. Replace the flanges if these conditions are present. Inspect the seal contact area in the case for metal burrs and gouges. Remove metal burrs with a fine-toothed file. Prior to seal installation, lube the seal lips, and apply sealer to the outer diameter of the seal case. Inspect the remote venting system on the transfer case and differentials for restrictions, kinks, and the proper location to ensure that pressure will not build up and that water should not enter the transfer case.

Question #70

Answer A is correct. If the differential case/carrier runout is excessive, the ring gear runout will be excessive.

Answer B is wrong. Side-bearing preload does not affect ring gear runout.

Answer C is wrong. Ring gear runout has no relation to the side gear end play.

Answer D is wrong. This is a good example of a question that may be the correct answer if you over think it. Since excessive torque could cause a differential case/carrier to warp, it could be the cause of the differential case/carrier runout, but not the root problem.

To determine if excessive ring rear runout is caused by the ring gear on the case/carrier, remove the ring gear and case/carrier and remove the ring gear from the case/carrier. Install the case/carrier assembly without the ring gear, and be sure the side bearings are in good condition and properly torqued. Position the dial indicator against the backside of the case/carrier, and rotate the case/carrier one revolution to measure the case/carrier runout. If the case/carrier runout is normal but the ring gear runout is excessive, replace the ring gear and pinion. When the case/carrier runout is excessive, replace the case/carrier.

Question #71

Answer A is wrong. Hard shifting would be caused only if the clutch pedal free play were excessive.

Answer B is wrong. Incomplete clutch release would be caused by excessive pedal free play.

Answer C is wrong. Improper clutch pedal free play cannot cause transaxle gear damage. On a clutch with an adjustable linkage, the release bearing should not be in contact with the pressure plate fingers. If the release bearing is not touching the fingers, it will not make any noise even if the bearing is bad. Clutch pedal free play is the distance between the release bearing and the pressure plate fingers. It is the gap or movement in the clutch pedal before the release bearing contacts the pressure plate and releases the clutch.

Hydraulic-controlled clutch systems use a release bearing that is always in contact with the fingers on the pressure plate. There is no manual adjustment on a hydraulic clutch system; it adjusts automatically as the clutch disc wears. When a clutch is disengaged, the release bearing moves towards the pressure plate. The release bearing continues to move towards the pressure plate fingers and compresses the springs in the pressure plate to release the clutch.

Answer D is correct. Lack of free play may keep the clutch from fully engaging.

Question #72

Answer A is wrong. The differential bearing preload does not affect axle flange runout.

Answer B is correct. A bent axle shaft will cause excessive flange runout.

Answer C is wrong. Pinion depth has nothing to do with axle flange runout.

Answer D is wrong. Limited slip clutch discs have nothing to do with axle flange runout.

The axle shaft runout may be measured by placing the axle shaft in a pair of "V" blocks and positioning a dial indicator against the center of the shaft. Rotate the shaft and observe the runout on the dial indicator. Shafts with excessive runout must be replaced.

Question #73
Answer A is wrong. Technician A is wrong. A collapsible pinion shaft spacer is not reusable and should be discarded after disassembly.
Answer B is wrong. Technician B is wrong; No pinion adjustment is made by loosening the nut once proper bearing preload has been established.
Answer C is wrong. Neither Technician is correct. Before pressing the new rear pinion bearing onto the pinion shaft, be sure the proper spacer washer is installed behind the pinion bearing. The shim determines the pinion depth. When the differential has been disassembled and overhauled, a new collapsible spacer and pinion nut must be installed. Tighten the pinion nut a small amount at a time until the specified turning torque is obtained with an inch-pound torque wrench and socket installed on the pinion nut. Never loosen the pinion nut.
Answer D is correct. Neither Technician is correct.

Question #74
Answer A is correct. Only Technician A is correct. Clutch pedal free play sets the distance between the release bearing and the pressure plate fingers.
Answer B is wrong. A worn release bearing makes more noise when the clutch pedal is depressed and the release bearing is in contact with the pressure plate.
Answer C is wrong. Only Technician A is correct.
Answer D is wrong. Only Technician A is correct. Some vehicles have linkages connected from the clutch pedal to the release fork. In other clutch systems, the clutch pedal is connected to the release fork by a cable. Another popular method designed to operate the clutch is a hydraulically controlled system. Many clutch linkages or cables have an adjustment to set the clutch pedal free play. Clutch pedal free play is the amount of pedal movement before the release bearing contacts the pressure plate release fingers or diaphragm. Many late-model vehicles have a self-adjusting clutch cable. In these clutches the cable is wrapped around and attached to a toothed wheel, and a ratcheting spring-loaded pawl is engaged with the toothed wheel. Each time the clutch pedal is released, the pawl removes any slack from the cable by engaging the next tooth on the wheel. The self-adjusting clutch cable system and the hydraulic clutch system have no built-in free play, and they use a constant running release bearing. The clutch release bearing is connected to the release fork, and the center opening in this bearing is mounted on a machined hub that is bolted to the front of the transmission.
Inspect clutch cable systems for binding, stretching cables and firewall flexing. These conditions prevent the clutch from fully releasing and will cause gear clash. Inspect hydraulic clutch systems for leaks and air in the system, which cause gear-clash concerns. Measuring throw-out bearing travel will determine if the clutch is not fully releasing and is causing the gear-clash concerns. This can only be performed on transmissions that have an inspection cover allowing you to view the throw-out bearing.

Question #75
Answer A is correct. Only Technician A is correct. The collapsible spacer must always be replaced.
Answer B is wrong. Pinion depth is adjusted by installing shims onto the pinion mountings. There is no such thing as a "selective race."
Answer C is wrong. Only Technician A is correct.
Answer D is wrong. Only Technician A is correct. If you are not familiar with the process of rebuilding differential, it would be wise to read through any manufacturer's procedure, as it exceeds the scope of this material. Key issues that you should know are:
The ring and pinion set must be replaced as a set; collapsible spacers must be replaced if they are over tightened or when the pinion bearings need to be replaced; and bearings should be replaced as assemblies.

Question #76
Answer A is wrong. Technician A is wrong. The drive pinion should be moved away from the drive gear.
Answer B is wrong. Technician B is wrong. The drive pinion should be moved away from the drive gear.
Answer C is wrong. Neither Technician is correct. Observe the tooth contact pattern on the ring gear. This pattern should be centered on the drive side of the ring gear teeth. If the ring gear tooth contact pattern is not correct, the drive pinion has to be moved in relation to the ring gear, or the backlash has to be adjusted to provide the correct pattern.
Answer D is correct. Neither Technician is correct.

Question #77
Answer A is wrong. Axle bearings have no relation to the ring-and-pinion contact pattern.
Answer B is correct. A worn collapsible spacer for the drive pinion will affect the ring-and-pinion contact pattern.
Answer C is wrong. Spider gears have no relation to the ring-and-pinion contact pattern.
Answer D is wrong. Differential bearings do not normally affect the ring-and-pinion contact pattern.
Prussian blue (a marking compound) is commonly used to check the tooth contact pattern on gear setup.
A used gear set will have a shiny pattern on the gear teeth that can be visually inspected. After assembly, look for the pattern to be centered with either new or used gears. Carrier-bearing adjustment may be necessary to correct the depth and backlash of the gears. Do not concern yourself with the exact appearance of the pattern as it varies by gear vendor. Take some time to look at the location of the pattern on the ring and pinion faces for your favorite manufacturer.

Question #78
Answer A is wrong. Drive pinion bearings would not cause a vibration while turning. They would cause noise all the time.
Answer B is correct. Only Technician B is correct. Differential side bearings are under load only during a turn.
Answer C is wrong. Only Technician B is correct.
Answer D is wrong. Only Technician B is correct. Inspection and side-play measurement of the side gear, pinion gears, thrust washer, and case/carrier in a rear-wheel-drive differential is basically the same as the inspection of these components in a transaxle differential.
The side/axle gear teeth and shims should be inspected for chips, pits, and cracks. Check the gear and housing bore for roughness and scoring. The pinion shaft and bearings must be inspected for roughness, scoring, and pitting. Replacement is required for any of these conditions.
Reassemble the differential side gears, pinion gears, pinion gear thrust washers, and shaft. Do not install the side gear thrust washers. Install the roll pin/bolt to retain this shaft in the differential case/carrier. Rotate the side gears two revolutions in each direction. Install a special tool through one of the axle openings against the side gear, and assemble a dial indicator so the stem rests against the tool. Move the side gear up and down and record the end play on the dial indicator. Rotate the side gear 90 degrees and record another end play reading. Turn the side gear another 90 degrees and record a third end play reading.
Use the smallest end play recorded to calculate the required side gear washer thickness. The side gear end play must be 0.001 in. to 0-013 in. If the end play recorded was 0.050 in., install a 0.042 in. washer, which provides an end play of 0.008 in. Once the side gear end play is determined to be correct, lubricate all parts and shims, and reassemble the unit.

Question #79
Answer A is wrong. The front driveshaft must be disconnected from the transfer case to remove the case.
Answer B is correct. Most transfer cases can be removed without removing the vehicle transmission.
Answer C is wrong. The rear driveshaft must be disconnected from the transfer case to remove the case.
Answer D is wrong. Linkage must be disconnected from the transfer case.
Disconnect the negative battery cable. If the vehicle is equipped with an air bag, wait for the time period specified by the vehicle manufacturer before working on the vehicle. Prior to transfer case removal, the vehicle should be raised on a lift. All safety precautions must be observed regarding lift operation and proper vehicle positioning on the lift. The transfer case drain plug should be removed and the lubricant drained into a suitable container. The front and rear driveshafts must be marked in relation to their matching flanges, followed by driveshaft removal. If the vehicle has a torsion bar front suspension, the torsion bars and rear torsions bar support may have to be removed to allow transfer case removal. Observe all safety precautions and proper service procedures regarding torsion bar removal. Remove the transfer case skid plate, shift linkage, and all electrical connectors. Support the transfer case on a high-lift jack, and use a safety strap to secure the transfer case on the jack. Remove the transfer case-to-transmission mounting bolts, and lower the transfer case with the jack.

Question #80
Answer A is wrong. A worn clutch release bearing makes a squeaking or rattling noise when the clutch pedal is depressed.
Answer B is wrong. A worn pilot bearing would not likely cause clutch chatter, but would cause shifting problems.
Answer C is wrong. Excessive input shaft play would cause noises and shifting problems, but not likely clutch chatter. A straightedge and a feeler gauge should be used to measure pressure plate warpage. A warped flywheel or pressure plate will cause clutch pulsations and chatter. If the flywheel does not have locating dowels, always punch mark the pressure plate and flywheel to ensure the pressure plate is reinstalled in the original position. The will reduce the chances of a vibration being induced into the clutch system.
When the clutch disc and pressure plate are installed on the flywheel, the clutch disc must be installed in the proper position. Install a clutch disc alignment tool though the clutch disc hub into the pilot bearing/bushing to be sure the clutch is properly aligned before attempting to reinstall the transmission/transaxle. The clutch disc area with the torsional springs must face away from the flywheel. An alignment tool is used to position the clutch disc and align it with the pilot bearing/bushing prior to tightening the pressure plate bolts.
Never use an impact wrench to tighten the pressure plate bolts because it may damage or weaken the fastener.
Answer D is correct. The torsional springs are designed to help reduce clutch chatter.

Question #81
Answer A is wrong. A dial indicator is used to measure differential case/carrier runout.
Answer B is correct. A dial indicator is used to measure differential case/carrier runout.
Answer C is wrong. A dial indicator is used to measure differential case/carrier runout.
Answer D is wrong. A dial indicator is used to measure differential case/carrier runout.
The ring gear runout should be measured prior to disassembling the differential. Mount the dial indicator assembly, and position the dial indicator stem at a 90-degree angle against the back of the ring gear. Turn the dial indicator to zero and rotate the ring gear one revolution. The difference between the highest and lowest dial indicator reading is the ring gear runout.
To determine if excessive ring rear runout is caused by the ring gear on the case, remove the ring gear and case and remove the ring gear from the case. Install the case assembly without the ring gear, and be sure the side bearings are in good condition and properly torqued. Position the dial indicator against the backside of the case, and rotate the case one revolution to measure the case runout. If the case runout is normal but the ring gear runout is excessive, replace the ring gear and pinion. When the case runout is excessive, replace the case.

Question #82
Answer A is correct. Too much driveline angle would cause the U-joints to bind.
Answer B is wrong. Gear problems or tire problems would likely cause a loud humming noise.
Answer C is wrong. The pinion gear is not as likely to be affected by an excessive driveline angle.
Answer D is wrong. Transmission mount damage is not as likely to result from an excessive driveline angle.
Driveshaft working angle is also known as pinion angle. The engine and transmission are installed in the chassis at a preset angle, usually pointing down at the tailshaft in rear-wheel drive applications. For our purposes, let's assume that the tailshaft is pointing down 3°. To avoid a humming or droning vibration in the driveline, the pinion must be pointing up 3°. This creates a situation where the planes in which they operate are parallel, providing the smoothest U-joint operation. (Note that most manufacturers give these measurement specifications only for pick-up trucks. Unless the manufacturer specifies differently, the static setting of each should be the same.) Causes of driveshaft working angle problems are: sagging (damaged) rear springs, sagging engine or transmission mounts, or major changes in ride height up or down.

Question #83

Answer A is wrong. Technician B is also correct. The friction plates wear and can develop hotspots with use.
Answer B is wrong. Technician A is also correct. The plates require a special lubricant to keep from chattering.
Answer C is correct. Both Technicians are correct.
Answer D is wrong. Both Technicians are correct. Limited slip differentials require a special lubricant specified by the vehicle manufacturer. The use of improper lubricant in a limited slip differential may result in differential noise, such as chattering while cornering. Clicking while turning a corner may be caused by worn limited slip components, such as clutches, preload springs, and shims. Limited slip differentials should be inspected for fiber and metal cuttings in the bottom of the housing. With the differential components removed, these cuttings may be flushed out of the housing with a solvent gun and an approved cleaning solution. Wipe the bottom of the housing out with a clean shop towel. Always use new gaskets or sealer when the differential and the cover are installed in the housing. Fill the differential to the bottom of the filler plug opening with the vehicle manufacture's specified limited slip differential lubricant.

Question #84

Answer A is correct. The linkage alignment pin can only be installed in neutral.
Answer B is wrong. The adjustment must be made with no load on the linkage.
Answer C is wrong. The adjustment must be made with no load on the linkage.
Answer D is wrong. The adjustment must be made with no load on the linkage.
Most external shift linkages and cables require adjusting, and a similar adjustment procedure is used on some vehicles. Raise the vehicle on a lift and place the shift lever in neutral to begin the shift linkage adjustment. With a lever-type shift linkage install a rod in the adjustment hole in the shifter assembly.
Adjust the shift linkage by loosening the rod-retaining locknuts and move the levers until the rod fully enters the alignment holes. Tighten the locknuts and check the shift operation in all gears. Rod-type shift linkages are adjusted with basically the same procedure as the lever-type linkages. When the alignment pin is in place, adjust the shift rod so the pin slides freely in and out of the alignment hole.

Question #85

Answer A is wrong. Friction plate thickness is measured with a micrometer.
Answer B is wrong. Steel plate thickness is measured with a micrometer.
Answer C is correct. A feeler gauge can be used to determine the required shim thickness.
Answer D is wrong. Spring tension is not measured with a feeler gauge.
Limited slip differentials have a set of multiple disc clutches behind each side gear to control differential action. The steel plates in each clutch set are splined to the case/carrier, and the friction placed between the steel placed are splined to the side gear clutch hub. Each clutch set has a preload spring that applies initial force to the clutch packs. A steel shim located in each clutch set controls preload. A friction plate always is placed next to the hub. Clicking while turning a corner may be caused by worn limited slip components, such as clutches, preload springs, and shims. Limited slip differentials should be inspected for fiber and metal cuttings in the bottom of the housing.

Question #86

Answer A is wrong. If the vehicle's U-joints were worn to the point of causing a vibration, it would not only be evident while cornering, but at all speeds.
Answer B is correct. Only Technician B is correct. Outboard U-joint vibration increases during a turn.
Answer C is wrong. Only Technician B is correct.
Answer D is wrong. Only Technician B is correct. Many AWD and 4WD front axle joints have an inner tripod joint. A snap-ring holds the inner spider assembly onto the shaft in this type of joint. Remove the outer boot clamps and slide the boot toward the center of the axle. Spread the retaining ring with a pair of snap-ring pliers, and then slide the outer CV joint from the axle shaft.
Inspect the inner and outer joint components for looseness, wear, scoring, and damage. Check the axle boots for cracks, oil soaking, and deterioration. Replace the boot and axle joints as required. Always install all of the grease supplied with the replacement joints in the joint. Always install the boot clamps using the manufacturer's recommended procedure. Some outer boot clamps must be installed with a special swaging tool.

Question #87

Answer A is wrong. Technician B is also correct. The switch closes when the transmission is shifted into reverse and usually receives battery voltage with the ignition in RUN.

Answer B is wrong. Technician A is also correct. The backup lamp switch is normally open and closes when the transmission is placed in reverse.

Answer C is correct. Both Technicians are correct.

Answer D is wrong. Both Technicians are correct. Manual transmissions may have various sensors, switches, and solenoids. These components may include a vehicle speed sensor (VSS), a backup light switch, and a computer-aided gear-select solenoid. The VSS sends a voltage signal to the PCM in relation to vehicle speed. The PCM uses this signal to operate output such as the cruise control. The backup light switch is operated by linkage inside the transmissions to operate the backup lights when the transmission is placed in reverse. The computer-aided gear-select solenoid ensures good fuel economy and compliance with federal fuel economy standards by inhibiting second and third gears when shifting out of first gear, under certain conditions.

Question #88

Answer A is wrong. Technician B is also correct. If the vent is left on top of the transfer case, water may enter through the vent.

Answer B is wrong. Technician A is also correct. As the transfer case warms up while the vehicle is being driven; pressure builds up in the case, which can force lubricant out at the weakest point in a seal or gasket.

Answer C is correct. Both Technicians are correct.

Answer D is wrong. Both Technicians are correct. Inspect the front and rear output shaft flanges in the transfer case for evidence of oil leakage in the seal area. Drain the fluid to check for water contamination. If there is any sign of leakage at the driveshaft flange seals, these seals must be replaced. When the flanges and seals are removed, always inspect the seal contact area on the flanges for roughness and scoring. Replace the flanges if these conditions are present.
Inspect the seal contact area in the case for metal burrs and gouges. Remove metal burrs with a fine-toothed file. Prior to seal installation, lube the seal lips, and apply sealer to the outer diameter of the seal case. Inspect the remote venting system on the transfer case and differentials for restrictions, kinks, and the proper location to ensure that pressure will not build up and that water should not enter the transfer case.

Question #89

Answer A is wrong. A worn speedometer drive and driven gears would not cause premature bushing wear.

Answer B is correct. Uneven mating surfaces can offset the extension housing and cause uneven bushing wear.

Answer C is wrong. A plugged transmission vent opening would cause the fluid to overheat and overflow.

Answer D is wrong. Main shaft end play does not usually cause bushing wear.
Inspect the extension housing for cracks, and check mating surfaces of this housing and the transmission case for metal burrs and gouges. Remove any metal burrs with a fine-toothed file. Replace the extension housing seal and check the extension housing bushing for wear or damage.

Answers to the Test Questions for the Additional Test Questions Section 6

1.	D	30.	A	59.	B	88.	D
2.	D	31.	A	60.	D	89.	D
3.	A	32.	C	61.	C	90.	D
4.	D	33.	D	62.	A	91.	C
5.	C	34.	A	63.	A	92.	B
6.	A	35.	C	64.	B	93.	C
7.	D	36.	B	65.	B	94.	D
8.	B	37.	C	66.	B	95.	A
9.	B	38.	C	67.	C	96.	A
10.	B	39.	A	68.	C	97.	C
11.	C	40.	C	69.	A	98.	B
12.	A	41.	C	70.	C	99.	C
13.	D	42.	B	71.	C	100.	D
14.	D	43.	A	72.	D	101.	C
15.	D	44.	A	73.	A	102.	C
16.	C	45.	D	74.	D	103.	D
17.	C	46.	B	75.	A	104.	B
18.	A	47.	D	76.	B	105.	C
19.	A	48.	D	77.	C	106.	C
20.	C	49.	B	78.	C	107.	B
21.	C	50.	C	79.	D	108.	A
22.	A	51.	C	80.	C	109.	D
23.	A	52.	D	81.	A	110.	A
24.	C	53.	B	82.	C	111.	B
25.	A	54.	D	83.	C	112.	A
26.	D	55.	C	84.	A	113.	D
27.	A	56.	A	85.	C	114.	C
28.	C	57.	C	86.	D	115.	B
29.	D	58.	C	87.	C	116.	C

117.	C	120.	D	123.	B
118.	D	121.	C	124.	C
119.	C	122.	D	125.	D

Explanations to the Answers for the Additional Test Questions Section 6

Question #1
Answer A is wrong. This could have a direct effect on the seal and sealing surface.
Answer B is wrong. This could have a direct effect on the seal and sealing surface.
Answer C is wrong. This could have a direct effect on the seal and sealing surface.
Many axle shafts ride directly on the axle bearing and seem to be more susceptible to wear, which results in noise and fluid leaks. Noises in differentials can be from the axle bearings, ring and pinion, any of the differential bearings, or differential side gears. Axle bearings usually can be isolated by the change in noise as the vehicle experiences different side loads or by raising the vehicle and running it in gear. Most axle bearing noises will subside dramatically when weight is taken off the wheels. Ring and pinion noises are associated with a whine or growl that changes in pitch as vehicle speed or engine load changes. To diagnose differential-carrier and pinion-bearing noises, a stethoscope is often employed to pinpoint the location of the noise.
Answer D is correct. The brake drum does not have a direct effect on the seal and sealing surface.

Question #2
Answer A is wrong. Worn axle shaft bearings may cause axle shaft seal leakage.
Answer B is wrong. A plugged axle vent may cause axle shaft seal leakage.
Answer C is wrong. Scored axle shafts may damage the seal, which could lead to a leak.
A worn axle shaft bearing can cause axle shaft seal failure. When the bearings wear out, axle side movement applies a greater load on the seal lip. Scored axle shafts in the seal area will damage the seal lip and cause the seal to fail. The axle shaft seal comes with a sprayed on sealant and does not require any other sealant. The sealing lip of the axle shaft seal should be lubricated with a light coating of gear lube to prolong the seal life.
If the axle shaft is damaged near the seal area, the shaft should be discarded and replaced with a new one.
Answer D is correct. The seal is made of heat-treated rubber to resist heat damage.

Question #3
Answer A is correct. An axle shaft should be stored standing up.
Answer B is wrong. Damage to the splines can occur this way.
Answer C is wrong. Runout cannot be checked in this way.
Answer D is wrong. If the axle is standing up and someone lowers the car, a disaster could occur.
A worn axle shaft bearing can cause axle shaft seal failure. When the bearings wear out, axle side movement applies a greater load on the seal lip. Scored axle shafts in the seal area will damage the seal lip and cause the seal to fail. The axle shaft seal comes with a sprayed on sealant and does not require any other sealant. The sealing lip of the axle shaft seal should be lubricated with a light coating of gear lube to prolong the seal life. If the axle shaft is damaged near the seal area, the shaft should be discarded and replaced with a new one.

Question #4
Answer A is wrong. Bronze bushings would not be used on an input shaft.
Answer B is wrong. Needle bearings could not support the weight of the input shaft.
Answer C is wrong. The input bearing is lubricated by transmission fluid.
An input shaft only uses one type of bearing: a ball bearing located toward the front half of the shaft (normally a pressed fit to the shaft). The bearing is lubricated by the fluid in the transmission. A needle bearing or a bronze bushing would not support the load that an input shaft is subjected to.
Answer D is correct. A ball bearing assembly is commonly used on an input shaft.

Question #5
Answer A is wrong. Technician B is also correct. If the vacuum motor does not operate, the 4WD system will not engage.
Answer B is wrong. Technician A is also correct. The vacuum motor needs a good vacuum source to operate.
Answer C is correct. Both Technicians are correct.
Answer D is wrong. Both Technicians are correct. Four-wheel-drive vehicles may have manual or automatic locking hubs. After the wheel bearings have been adjusted and the locking screws or nuts tightened to the specified torque, place some multi-purpose grease on the hub inner spines. Do not pack the hub with grease. Slide the locking hub assembly into the wheel hub until it seats. Install the lock ring in the wheel hub groove. Install the lock washer and axle shaft stop on the axle bolt; then install the bolt and tighten it to the specified torque. Place a small amount of lubricant on the cap seal, and install the cap over the body and into the wheel hub. Install the attaching bolts and tighten them to the specified torque. Rotate the locking hub control from stop to stop to make sure it operates freely. Set both controls to the same auto or lock position. Automatic locking hubs may operate on engine vacuum. These hubs operate by an electrical signal from the GEM module to a solenoid that controls the vacuum to the actuator hubs. The weakness of these systems is that a small vacuum leak may prevent the hubs from locking. Perform testing by plugging the supply end of the system and applying vacuum with at the other end with a hand vacuum pump. Then observe the gauge to determine if a leak exists. Also make sure that a sufficient supply of vacuum is available to the control solenoid and that the solenoid functions

Question #6
Answer A is correct. Only Technician A is correct. An oil-soaked transmission mount should be replaced.
Answer B is wrong. An oil-soaked transmission mount should be replaced.
Answer C is wrong. Only Technician A is correct.
Answer D is wrong. Only Technician A is correct. Engine and transmission mounts should be inspected for broken, sagged, oil-soaked, or deteriorated conditions.
Any of these mount conditions may cause a grabbing, binding clutch. On a rear-wheel drive car, damaged engine or transmission mounts may cause improper driveshaft angles, which result in a vibration that changes in intensity when the vehicle accelerates and decelerates.

Question #7
Answer A is wrong. A bad electronic shift motor may cause the transfer case to not engage.
Answer B is wrong. A blown fuse may cause the transfer case to not engage.
Answer C is wrong. A bad 4WD engage switch may cause the transfer case to not engage.
An electronic-shift transfer case has all of the shift linkage inside the case. If an electronic-shift transfer case has a problem with shifting, an electrical component would be a likely cause of failure.
Answer D is correct. An electronic transfer case does not have any linkage exposed that can rust.

Question #8
Answer A is wrong. The procedure does not indicate shift cable removal.
Answer B is correct. The shift cable is adjusted in the neutral position.
Answer C is wrong. Shift cables should never be modified.
Answer D is wrong. The transaxle case is not being repaired.
A misadjusted shift linkage may cause many problems. If the linkage is misadjusted, the transmission may not be able to be shifted all the way into gear. This will cause further damage to the transmission. An improper shift linkage adjustment also may cause hard shifting or sticking in gear.
Most transmissions and transaxles are adjusted with the unit in neutral. A 1/4-inch (6.35 mm) bar or drill bit is installed in the lever to hold the transaxle in neutral while the cables or linkage are adjusted. Transmissions and transaxles with internal linkage have no adjustment. Shift cables should not be modified in any way—only replaced or adjusted.

Question #9
Answer A is wrong. Ring gear runout is not being measured.
Answer B is correct. Ring gear backlash is being measured.
Answer C is wrong. Ring gear backlash is being measured.
Answer D is wrong. Bearing preload is not being measured.
Ring gear runout and case side play should be measured before removing the ring gear and case assembly.
The side bearing caps should be marked in relation to the case before removal. The side bearings should be lubricated before installation.
The side bearings must be in good condition before measuring case runout. The ring gear runout should be measured before the case runout.

Question #10
Answer A is wrong. Replacing the engine mounts would not affect bell housing bore alignment.
Answer B is correct. Adjustable bell housing dowels will correct the runout problem.
Answer C is wrong. Bell housing shims would not correct a bell housing bore alignment condition.
Answer D is wrong. The clutch disc does not control bell housing bore alignment.
When bell housing face squareness is measured, a dial indicator is attached to a special tool mounted in the center of the clutch disc splined opening, with the dial indicator stem contacting the bell housing face.
The tool is rotated to obtain the measurement. If the bell housing runout is excessive, shims may be installed between the bell housing and engine block to correct this misalignment.
When performing the bell housing bore runout measurement, a dial indicator is attached to a special tool mounted in the center of the clutch disc splined opening, with the dial indicator stem contacting the bell housing bore. The tool is rotated to obtain the measurement. Some engines have offset dowel pins in the block surface that mates with the bell housing surface. These offset dowel pins may be rotated to correct the bell housing runout. If the offset dowel adjustment is not enough to correct the bell housing bore runout, replace the bell housing.

Question #11
Answer A is wrong. Worn friction plates may cause the limited slip differential to not function properly.
Answer B is wrong. Weak spring tension could cause a limited slip differential to fail.
Answer C is correct. The friction plates are not under excessive load and are unlikely to strip the teeth.
Answer D is wrong. The wrong fluid in the differential could cause a limited slip differential to fail.
Limited slip differentials have a set of multiple disc clutches behind each side gear to control differential action. The steel plates in each clutch set are splined to the case, and the friction plates between the steel placed are splined to the side gear clutch hub. Each clutch set has a preload spring that applies initial force to the clutch packs. A steel shim in each clutch set controls preload. A friction plate always is placed next to the hub.

Question #12
Answer A is correct. Only Technician A is correct. Sharp edges help to engage the synchronizer.
Answer B is wrong. The sharp edges will not prevent the transmission from jumping out of gear.
Answer C is wrong. Only Technician A is correct.
Answer D is wrong. Only Technician A is correct. The blocking ring dogteeth tips should be pointed with smooth surfaces.
Clearance between the blocking ring and the matching gear dogteeth is important for proper shifting. The synchronizer sleeve must slide freely on the synchronizer hub. The threads on the blocking ring in the cone area must be sharp to get a good bite on the gear to stop it from spinning and to make a synchronized non-clashing shift.
If the clearance between the blocking ring and the fourth-speed gear dogteeth is less than specified, the blocking ring is worn, which results in hard shifting. This problem would not result in noise while driving in fourth gear.

Question #13
Answer A is wrong. Transfer cases do not have sight glasses.
Answer B is wrong. Transfer cases do not have fill vents.
Answer C is wrong. Neither Technician is correct. When filling a transmission or differential case/carrier, fluid will come out of the fill hole as it reaches the full level. After draining differential fluid, inspect it for excessive metal particles. Silver- or steel-colored particles are signs of gear or bearing wear. Copper- or bronze-colored particles are signs of limited slip clutch disc wear.
Answer D is correct. Neither Technician is correct.

Question #14
Answer A is wrong. Over time, the transfer case chain stretches.
Answer B is wrong. Select-fit thrust washers should be measured and used for assembly of the transfer case.
Answer C is wrong. During disassembly, the transfer case components should be thoroughly inspected.
Prior to disassembly, the transfer case should be cleaned externally with a suitable cleaning solution and a brush. The transfer case must be properly secured in a bench-mounted holding fixture. The disassembly and reassembly procedures vary, depending on the transfer case manufacturer. Always follow the procedure in the vehicle manufacturer's service manual. The following is a general disassembly procedure:
Remove all electrical sensors, switches, and motors from the transfer case. Remove the rear extension housing, followed by the speedometer spacer, drive gear, and steer ball. Remove the rear output shaft snap-ring, and the transfer case to cover bolts. Separate the transfer case cover from the case. Remove, clean, and inspect all transfer case components.
Use a clean shop towel to clean the magnet in the transfer case. An oil pump is mounted in many transfer cases to circulate lubricant to all the case components. The oil pump is sometimes driven by the output shaft. In some transfer cases, the oil pump is not repairable. Use the proper seal drivers to replace all transfer case seals. Replace all worn or damaged transfer case components. Use compressed air to blow out all oil passages in the case or case components. When reassembling the transfer case, some manufacturers recommend sealing the case to the cover with a small bead of black silicone rubber sealant. An excessive amount of sealant may contaminate the transfer case fluid and plug the oil pump pickup screen, resulting in case component failure. After the transfer case is properly assembled and installed, it must be filled to the specified level with the manufacturer's specified lubricant. If the wrong fluid is placed in the transfer case, hard shifting and component damage may occur.
Answer D is correct. All the transfer case parts should be cleaned then lubricated before assembly.

Question #15
Answer A is wrong. Driveshaft runout is checked with a dial indicator positioned near the center of the driveshaft.
Answer B is wrong. Driveshaft runout is checked with a dial indicator positioned near the center of the driveshaft.
Answer C is wrong. Neither Technician is correct. While measuring driveshaft runout, the dial indicator should be positioned near the center of the driveshaft.
The driveshaft should be replaced if the runout is excessive. A driveshaft that is bent or damaged in any way should be replaced; repairs to a damaged driveshaft should not be attempted. To check driveshaft runout, the vehicle should be raised and a dial indicator with a magnetic base should be installed under the vehicle near the center of the driveshaft.
The surface of the shaft should be wiped off or cleaned in case it is rusted or has dirt buildup that may affect the reading on the dial indicator. Rolling a driveshaft on a flat surface is not an accurate way of checking the driveshaft runout.
Answer D is correct. Neither Technician is correct.

Question #16
Answer A is wrong. Technician B is also correct. Many modern blocking rings have special composite layers that can be damaged by the incorrect fluid type.
Answer B is wrong. Technician A is also correct. Improper fluid type can cause manual transmission shift problems; this is why it is so important the fluid be of manufacturer's recommended type and viscosity.
Answer C is correct. Both Technicians are correct.
Answer D is wrong. Both Technicians are correct. It is important for technicians to include an inspection of drivetrain fluid level and to be sure that manufacturers recommended types of fluids be used. As a technician, you should check manufacturer's TSBs for changes in fluids or additives that have been recommended since the production date. When fluid level is incorrect, it should be adjusted with the correct fluid type to proper levels.
Some vehicle manufacturers recommend a mineral oil with an extreme pressure (EP) additive for manual transmissions. The most common gear oil classifications are SAE 75W, 75W-80, 80W-90, 85W-90, 90, or 140. Thicker gear oils have higher classification numbers. Some manual transmissions require engine oil or automatic transmission fluid. Many manual transaxles require 30W engine oil, 90W gear oil or ATF.
Using the wrong fluid can cause shifting and/or wear concerns. Using too thick oil in a transmission that is designed for 30W oil or ATF can cause shifting concerns during cold weather.

Question #17
Answer A is wrong. Technician B is also correct. The bushings and grommets may have become worn and need to be replaced.
Answer B is wrong. Technician A is also correct. The cable is the shift linkage and may have stretched or frayed.
Answer C is correct. Both Technicians are correct.
Answer D is wrong. Both Technicians are correct. A misadjusted shift linkage may cause many problems. If the linkage is misadjusted, the transmission may not be able to be shifted all the way into gear. This will cause further damage to the transmission. An improper shift linkage adjustment also may cause hard shifting or sticking in gear.
Most transmissions and transaxles are adjusted with the unit in neutral. A 1/4-inch (6.35 mm) bar or drill bit is installed in the lever to hold the transaxle in neutral while the cables or linkage are adjusted. Transmissions and transaxles with internal linkage have no adjustment. Shift cables should not be modified in any way, only replaced or adjusted.

Question #18
Answer A is correct. The input bearing retainer and seal work in conjunction with the input bearing.
Answer B is wrong. The countershaft bearing is not the Most-Likely component to be damaged.
Answer C is wrong. Second gear is not the Most-Likely component to be damaged.
Answer D is wrong. The counter gear is not the Most-Likely component to be damaged.
Some front bearing retainers have a seal and a shim behind the retainer. Remove the input gear and shaft and the output shaft assembly. Remove the countershaft and the countershaft/cluster gear assembly, followed by the reverse idle shaft and the reverse idler gear. Be sure the input shaft teeth that mesh with the synchronizer collar are not worn or chipped. Remove the roller bearings from the inner end of the input shaft. These bearings and the bearing contact area in the input shaft must be inspected for wear, pitting, and roughness. Check the input shaft bearing for smooth rotation and looseness.

Question #19
Answer A is correct. Only Technician A is correct. A worn blocker ring will cause gear clash.
Answer B is wrong. A worn blocker ring would have dull edges in the cone area.
Answer C is wrong. Only Technician A is correct.
Answer D is wrong. Only Technician A is correct. A worn blocker ring will cause the transmission/transaxle to have gear clash. If the blocker ring is worn in the cone area (meaning that all of the sharp ridges are dull or gone), the blocker ring will not work properly.
A blocker ring should stop a gear from spinning through the sharp ridges in the cone area before the synchronizer sleeve engages the gear.

Question #20
Answer A is wrong. Packing a bearing before installation is to pack the bearing with grease.
Answer B is wrong. Load is not related to preload.
Answer C is correct. Bearing preload is the amount of pressure applied to a bearing upon assembly.
Answer D is wrong. Bearing preload is adjustable.
Bearing preload is determined by selective shim thickness in some manual transmissions/transaxles and differentials.
Preload measurement is a critical procedure when replacing any "hard part" such as a shaft, bearing, case. or thrust washers
and shims. Reduced bearing preload may result in abnormal clicking, clunking noise and possible wear in components.
Excessive bearing preload may result in overheated, worn, and severely damaged bearings and components.

Question #21
Answer A is wrong. The transmission should be in neutral when checking driveshaft runout.
Answer B is wrong. A magnetic-base dial indicator should be installed near the center of the driveshaft.
Answer C is correct. This is not an acceptable way of inspecting a driveshaft.
Answer D is wrong. The driveshaft surface must be free of debris to ensure an accurate reading.
A dial indicator must be mounted to the vehicle underbody near the center of the driveshaft to measure driveshaft runout.
Position the dial indicator stem against the surface of the driveshaft. Be sure the driveshaft surface is clean and undamaged.
Rotate the driveshaft one revolution to measure the runout. When the runout exceeds specifications, replace the driveshaft
and recheck the runout. If the runout is still excessive, check for a bent U-joint flange or slip yoke.

Question #22
Answer A is correct. Only Technician A is correct. The center bore of the idler gear should be smooth and mar free.
Answer B is wrong. The reverse idler gear rides on needle bearings and is not splined in the center bore.
Answer C is wrong. Only Technician A is correct.
Answer D is wrong. Only Technician A is correct. The reverse idler gear teeth should be inspected for chips, pits, and
cracks. Check the gear bore for roughness and scoring. The needle bearings/bushings and shafts must be inspected
for roughness, scoring, and pitting.

Question #23
Answer A is correct. Only Technician A is correct. A cluster/counter gear must be replaced if one of the gears has a
damaged tooth.
Answer B is wrong. A damaged gear must always be replaced.
Answer C is wrong. Only Technician A is correct.
Answer D is wrong. Only Technician A is correct. Since the reverse idler gear is only in mesh with reverse gear, this
gear rotates in reverse gear only. Inspect the reverse idler gear for pitted, cracked, nicked, or broken teeth. Check its
center bore for a smooth surface.
Carefully inspect the needle bearings on which the idler gear rides for wear, burrs, and other defects.
Also, inspect the reverse idler gear's shaft surface for scoring, wear, and other imperfections. Replace any part that
is damaged or excessively worn.

Question #24
Answer A is wrong. Technician B is also correct. A broken cable may catch and slip repeatedly; or a bad head may be
sticking or slipping.
Answer B is wrong. Technician A is also correct. The gear may be worn and not be making a good contact with the
gear on the output shaft.
Answer C is correct. Both Technicians are correct.
Answer D is wrong. Both Technicians are correct. A worn speedometer drive and driven gears may cause erratic
speedometer operation, but this problem would not result in premature extension housing bushing failure.

Question #25
Answer A is correct. Only Technician A is correct. When the clutch is disengaged, the pilot bearing rotates on the input shaft, which allows the bearing to make noise.
Answer B is wrong. If the clutch is disengaged, the input shaft on the transmission is not spinning, eliminating the input shaft as a source.
Answer C is wrong. Only Technician A is correct.
Answer D is wrong. Only Technician A is correct. When the clutch is engaged, the transmission input shaft rotates at the same speed as the engine flywheel and pilot bearing at all times. When the clutch is disengaged, the flywheel and pilot bearing rotate on the end of the transmission input shaft and turn faster than the shaft.
If a bushing-type pilot bearing is lubricated with bearing grease, friction actually will increase between the bushing and the transmission input shaft. Lubricate a bushing-type pilot bearing with motor oil. Lubricate a roller-type pilot bearing with wheel bearing grease.

Question #26
Answer A is wrong. A bent linkage may prevent the transmission from engaging reverse.
Answer B is wrong. A broken shift fork may prevent the transmission from engaging reverse.
Answer C is wrong. A misadjusted linkage may prevent the transmission from engaging reverse.
Shift rails should be inspected to be sure that they are not bent. A bent shift rail will not cause the transmission to jump out of gear or result in a gear clash or a gear noise.
Hard shifting may be a cause of a bent shift rail. The shift rail is linked to the shifter handle; if the rail is bent, it may interfere with other parts in the transmission, causing the hard shift feeling when the shift lever is moved from certain gears. Common causes of a transmission that does not shift properly are both linkage that is bent or a linkage that is out of adjustment. Failure to go into gear is more commonly caused by a broken shift fork than by a damaged gear.
The shift linkages internal to the transmission are located at the top or side of the housing. Mounted inside the transmission is the control end of the shifter and the shift controls. The shift controls consist of the shift rail and the shift fork. As the shift fork moves toward the preferred gear, it moves the synchronizer sleeve to lock the speed gear to the shaft.
Answer D is correct. The transmission would engage in reverse, but may function poorly.

Question #27
Answer A is correct. Only Technician A is correct. A damaged release (throw-out) bearing noise will be more pronounced with the pedal depressed.
Answer B is wrong. A worn input shaft bearing will not cause a chirping sound with the clutch pedal depressed.
Answer C is wrong. Only Technician A is correct.
Answer D is wrong. Only Technician A is correct. Clutch chatter is felt normally when the pressure plate has made initial contact with the flywheel. Causes of clutch chatter, such as weak torsional springs, will not absorb the shock of the clutch when contact is made. Other causes of chatter could be a bent clutch disc or a burned or glazed lining on the disc. Clutch chatter may occur if the flywheel is glazed, has excessive runout, or is scored. If a clutch were disengaged, the input shaft on the transmission would not be spinning, thus eliminating the bearing noise coming from the input shaft bearing. A pilot bearing would be a probable cause of noise in this case. With the input shaft not spinning, the pilot bearing is spinning on the end of the input shaft. This could cause a bearing noise due to a bad or worn pilot bearing.

Question #28
Answer A is wrong. Technician B is also correct. Oil would contaminate the clutch disc and cause uneven clutch disc wear.
Answer B is wrong. Technician A is also correct. The clutch would not be making even contact with the flywheel.
Answer C is correct. Both Technicians are correct.
Answer D is wrong. Both Technicians are correct. Clutch chatter is felt normally when the pressure plate has made initial contact with the flywheel. Causes of clutch chatter, such as weak torsional springs, will not absorb the shock of the clutch when contact is made. Other causes of chatter could be a bent clutch disc or a burned or glazed lining on the disc. Clutch chatter may occur if the flywheel is glazed, has excessive runout, or is scored.

Question #29
Answer A is wrong. A short in the backup light switch may cause this problem.
Answer B is wrong. A misadjusted linkage could cause this problem.
Answer C is wrong. A short in the wiring harness may cause this problem.
The backup lamp switch is normally located in the transmission on a manual transmission-equipped vehicle.
The switch is normally open and has power going to it when the vehicle ignition is on. The backup lamp lights when the vehicle is shifted into reverse. The switch will close when the vehicle is shifted in reverse and provide a path to ground for the backup lamps to operate.
Answer D is correct. The brake light switch has no effect on the backup lights.

Question #30
Answer A is correct. Only Technician A is correct. Dry or sticking linkage can cause hard shifting.
Answer B is wrong. Too strong a pressure plate would not cause hard shifting.
Answer C is wrong. Only Technician A is correct.
Answer D is wrong. Only Technician A is correct. If the transmission shifts hard or the gears clash while shifting, a common cause of the problem is the clutch. Check the clutch pedal free-travel adjustment. Make sure the clutch releases completely. Also check for worn clutch parts and a binding input shaft pilot bearing. Shift linkage problems can also cause this problem. Proper engagement of gears is impossible if the shift lever is worn, binding, or out of adjustment. An unlubricated linkage will also cause shifting difficulties.

Question #31
Answer A is correct. A sensor should never be replaced as the first step after reading a trouble code.
Answer B is wrong. The harness could be tested as the next step after reading a trouble code.
Answer C is wrong. The service manual could be consulted as the next step after reading a trouble code.
Answer D is wrong. The connector could be inspected as the next step after reading a trouble code.
Manual transmissions may have various sensors, switches, and solenoids. These components may include a vehicle speed sensor (VSS), a backup light switch, and a computer-aided gear-select solenoid. The VSS sends a voltage signal to the PCM in relation to vehicle speed. The PCM uses this signal to operate output, such as the cruise control. Some vehicles multiplex sensors and obtain the VSS signal from ABS wheel speed sensors. The ABS/VSS sensor is usually a permanent-magnet signal generator. However, some units utilize a Hall-effect device to generate the voltage signal. While some of these types of sensors may be checked using an ohmmeter for initial resistance, the only accurate method of testing the sensors is by using a DMM on the AC voltage setting or by using an oscilloscope.

Question #32
Answer A is wrong. A new main shaft is not necessary to set end play.
Answer B is wrong. Bearings may be reused if they are inspected and in satisfactory condition.
Answer C is correct. Select-fit shims and thrust washers are used to set parts to specification.
Answer D is wrong. Snap-rings may be reused if they are not damaged, and snap-rings are not used to set this end play.
In fourth gear, the 1-2 synchronizer is moved ahead so the synchronizer hub is meshed with the dogteeth on the fourth-speed gear on the input shaft. Excessive input shaft end play would cause the transmission to jump out of fourth gear. The technician should be taking end play readings while disassembling a transmission; these readings will be recorded. When the transmission is reassembled, select-fit thrust washers and shims will be used to set all parts to specifications. A main shaft should not be replaced unless, during inspection, excessive wear or damage is found. Bearings and snap-rings can be reused if nothing is found during cleaning and inspection.

Question #33

Answer A is wrong. Measurements for tolerances and end play should be recorded and transferred to the transmission upon rebuilding it.

Answer B is wrong. All parts should be cleaned with solvent.

Answer C is wrong. Always pay attention to the condition of parts being removed; this may help in identifying the source of the problem.

Clean the transmission case with a steam cleaner, degreaser, or cleaning solvent. As you begin to disassemble the unit, pay close attention to the condition of its parts. Using a dial indicator, measure and record the end play of the input and main shafts. This information will be needed during assembly of the unit to select the appropriate selective shims and washers. Before assembly, the transmission parts should be cleaned with solvent.

Answer D is correct. Friction discs are not part of a manual transmission.

Question #34

Answer A is correct. Only Technician A is correct. Bearings and races are replaced as sets.

Answer B is wrong. Automatic locking hubs are NOT packed with grease prior to installation.

Answer C is wrong. Only Technician A is correct.

Answer D is wrong. Only Technician A is correct. Neither the automatic locking hubs nor the caps should be packed with grease. If they are packed with grease, they will not operate properly; the parts must move freely.

If only one bearing is bad, replace only that bearing. When a wheel bearing is replaced, the bearing race should also be replaced. The new bearing may fail if the race is not replaced.

Question #35

Answer A is wrong. Tightening both adjusters would make the preload tighter.

Answer B is wrong. Both adjusters must be changed at the same time.

Answer C is correct. This adjustment would increase the backlash.

Answer D is wrong. This adjustment would also decrease the backlash.

The backlash between the ring gear and pinion gear is adjusted after the differential is assembled. Mount a dial indicator on the differential housing and position the dial indicator stem against one of the ring gear teeth. Turn the dial indicator to zero and rock the ring gear back and forth against the pinion gear teeth. The ring gear backlash is indicated on the dial indicator. Vehicle manufacturers usually recommend measuring the backlash at several locations around the ring gear.

Side-bearing preload limits the amount of lateral differential case/carrier movement in the axle housing or carrier. When the differential has threaded adjusters on the outside of the side bearings, loosen the right adjuster and tighten the left adjuster to obtain zero backlash. Turn the right adjuster the specified amount to obtain the proper preload. Then rotate each adjuster the same amount in opposite directions to obtain the specified backlash.

Question #36

Answer A is wrong. The teeth should not be rounded.

Answer B is correct. Only Technician B is correct. The teeth should have a sharp, beveled edge.

Answer C is wrong. Only Technician B is correct.

Answer D is wrong. Only Technician B is correct. The blocking ring and all gear dogteeth tips should be pointed with smooth surfaces. For proper shifting, it is important to have clearance between the blocking ring and the matching gear dogteeth. The synchronizer sleeve must slide freely on the synchronizer hub. The threads on the blocking ring in the cone area must be sharp to get a good bite on the gear to stop it from spinning and to make a synchronized non-clashing shift. If the clearance between the blocking ring and the fourth-speed gear dogteeth is less than specified, the blocking ring is worn, which results in hard shifting. This problem would not result in noise while driving in fourth gear.

Question #37

Answer A is wrong. SAE 90 gear oil is not usually used in late-model manual transmissions.

Answer B is wrong. Power-steering fluid should not be used in a transmission.

Answer C is correct. Late-model manual transmissions use automatic transmission fluid or motor oil.

Answer D is wrong. Such a lubricant would be highly uncommon.

Transmission fluid level must be maintained at the level of the check plug in the transmission case or at a level marked on a transmission dipstick. Many late-model manual transmissions and transaxles use automatic transmission fluid (ATF) as a lubricant for reduced friction and improved vehicle fuel economy.

Some manual transmissions use hypoid gear oil as a lubricant; a few use motor oil. The hypoid ring-and-pinion gear sets in rear-wheel drive axles require hypoid gear oil, usually GL4 or GL5. Limited slip differentials require additional fluid additives. The viscosity of hypoid gear oil is higher (thicker) than that of motor oil or ATF. It may be single-viscosity, such as SAE 90, or multiple-viscosity, such as 85W-90. Many final-drive gear sets in front-wheel drive (FWD) transaxles are not hypoid gears, and they use ATF or motor oil as a lubricant. Some FWD final drives are hypoid gear sets, however, and require GL4 or GL5 gear oil.

Always follow the carmaker's specifications for fluid type, viscosity, and replacement intervals.

Question #38

Answer A is wrong. Technician B is also correct. The detent springs keep the shift rails from moving.

Answer B is wrong. Technician A is also correct. The synchronizer eliminates gear clash when shifting and keeps a transmission in gear.

Answer C is correct. Both Technicians A and B are correct.

Answer D is wrong. Both Technicians A and B are correct. Some vehicle manufacturers provide service procedures and specifications for measuring the end play on various transmission gears. In many end play adjustments, a dial indicator is mounted on the transmission case, and the dial indicator stem is positioned on the gear. Push the gear back and forth and observe the end play reading on the dial indicator. Excessive gear end play is usually caused by worn thrust washers. Inspect all gear thrust washers and retainers for wear.

Shift rails should be inspected to be sure they are not bent, broken, or worn. Inspect all the holes and notches in the shift rails for wear and damage. When each shift rail is installed in the appropriate bore in the case of the shift cover, check for excessive movement between the rail and the bore. Inspect all detent springs to be sure they are not worn, bent, or weak.

Question #39

Answer A is correct. Bearing surfaces are indicated in the figure.

Answer B is wrong. Gear journals are not indicated.

Answer C is wrong. Oil journals are not indicated.

Answer D is wrong. Synchronizer mounting locations are not indicated.

The main shaft is not drilled with oil journals. Inspect the bearing surfaces of the main shaft; it should be smooth and show no signs of overheating. Also inspect the gear journal areas on the shaft for roughness, scoring, and other defects. Check the shaft splines for wear, burrs, and other conditions that would interfere with the slip yoke's ability to slide smoothly on the splines.

Question #40
Answer A is wrong. Clutch throwout bearing would make noise when depressing the clutch pedal.
Answer B is wrong. Ring and pinion problems would also make noise while driving straight.
Answer C is correct. The differential side bearings would be the Most-Likely cause, since they are the only item that would experience any change during turning.
Answer D is wrong. Wrong lubricant would most likely cause noise in all gears.
A worn blocker ring, damaged speed gear, or a worn bushing could cause gear clash in a particular gear.
Gear clash in all gears could be caused by stretched shifter cables.
Damaged ring gear teeth would cause a clicking noise while the vehicle is in motion.
This problem would not cause differential chatter. Improper preload on differential components, such as side bearings, may cause differential chatter. If there were a constant whining noise coming from the differential, the noise could not be coming from the side gears or the spider gears. These gears are used only when the vehicle is turning; so if they were damaged, the noise would be heard only on turns. The wrong differential lube could cause damage to the differential parts, but it will not cause a whining noise. If the preload and backlash are not set properly, the gear mesh could be too tight and cause a whining noise.

Question #41
Answer A is wrong. Technician B is also correct. The floor- and column-mounted linkages can be adjusted.
Answer B is wrong. Technician A is also correct. There are no internal adjustments. The floor and column-mounted linkages can be adjusted, however.
Answer C is correct. Both Technicians are correct.
Answer D is wrong. Both Technicians are correct. Most external shift linkages and cables require adjusting, and a similar adjustment procedure is used on some vehicles. Raise the vehicle on a lift and place the shift lever in neutral to begin the shift linkage adjustment. With a lever-type shift linkage install a 1/4" rod in the adjustment hole in the shifter assembly. Adjust the shift linkage by loosening the rod-retaining locknuts and move the levers until the 1/4" rod fully enters the alignment holes. Tighten the locknuts and check the shift operation in al gears. Rodtype shift linkages are adjusted with basically the same procedure as the lever-type linkages. When the alignment pin is in place, adjust the shift rod so the pin slides freely in and out of the alignment hole.

Question #42
Answer A is wrong. Only worn bushings must be replaced.
Answer B is correct. Only Technician B is correct. Only worn bushings must be replaced.
Answer C is wrong. Only Technician B is correct.
Answer D is wrong. Only Technician B is correct. Most external shift linkages and cables require adjusting, and a similar adjustment procedure is used on some vehicles. Raise the vehicle and place the shifter in the neutral position to begin the shift linkage adjustment. With a lever-type shift linkage, install a rod in the adjustment hole in the shifter assembly. Adjust the shift linkages by loosening the rod retaining locknuts and moving the levers until the rod fully enters the alignment holes. Tighten the locknuts and check the shift operation in all gears.
Transmissions with internal linkage do not have provisions for adjustment. However, external linkages, both floor and column mount, can be adjusted. Linkages are adjusted at the factory, but worn parts may make adjustment necessary. Also, after a transmission has been disassembled, the shift lever may need adjustment.

Question #43

Answer A is correct. Only Technician A is correct. Constant running release bearings are used with hydraulically controlled clutches.

Answer B is wrong. The release bearings move toward the pressure plate to disengage the clutch.

Answer C is wrong. Only Technician A is correct.

Answer D is wrong. Only Technician A is correct. On a clutch with an adjustable linkage, the release bearing should not be in contact with the pressure plate fingers. If the release bearing is not touching the fingers, it will not make any noise even if the bearing is bad. Clutch pedal free play is the distance between the release bearing and the pressure plate fingers. It is the gap or movement in the clutch pedal before the release bearing contacts the pressure plate and releases the clutch. Hydraulically controlled clutch systems use a release bearing that is always in contact with the fingers on the pressure plate. There is no manual adjustment on a hydraulic clutch system; it adjusts automatically as the clutch disc wears. When a clutch is disengaged, the release bearing moves towards the pressure plate. The release bearing continues to move towards the pressure plate fingers, and it compresses the springs in the pressure plate to release the clutch

Question #44

Answer A is correct. Engine oil can contaminate the clutch disc.

Answer B is wrong. The engine should always be supported when the transmission is removed.

Answer C is wrong. Only Technician A is correct.

Answer D is wrong. Technician A is correct. Always verify the source of any fluid leak. While it may appear to be coming from the transmission, it may be a rear main engine oil leak.

If the transmission is supported from the input shaft, the weight of the transmission itself could cause the input shaft and the clutch disc to be bent or damaged. The engine support fixture must be installed before the transmission-to-engine bolts are even loosened. When the transmission is being removed, the clutch disc may move, causing it to be misaligned. A clutch disc alignment tool must be used before the transmission is installed to align the clutch disc with the flywheel.

Question #45

Answer A is wrong. Mounts saturated in oil should never be cleaned and reused.

Answer B is wrong. Mounts soaked in oil should always be replaced.

Answer C is wrong. Oil will cause the mounts to swell and weaken.

A transmission mount absorbs a lot of torque and vibration in the rubber mount when shifting and accelerating.

If the rubber mount becomes saturated with oil, the oil will deteriorate the rubber and weaken it; this will eventually cause the mount to fail. Oil will not cause the mount to crack, but the oil will make the rubber in the mount feel soft and spongy.

Answer D is correct. Mounts soaked in oil should always be replaced.

Question #46

Answer A is wrong. Input shaft end play should be checked before disassembling the transaxle.

Answer B is correct. Only Technician B is correct. Always rotate the input shaft to check turning effort before disassembling the transaxle.

Answer C is wrong. Only Technician B is correct.

Answer D is wrong. Only Technician B is correct. While assembling a manual transaxle, it is important to apply gear lube to all of the transaxle parts. Before checking the specifications of the shafts in the transaxle, rotate the shafts to work the gear lube into the bearings. If the gear lube is not worked into the bearings a false measurement may be made.

Before disassembling a transaxle, observe the effort it takes to rotate the input shaft through all forward gears and reverse. Extreme effort in any or all gears may indicate an end play problem or a bent shaft.

Question #47
Answer A is wrong. The axle shafts must be removed to remove the differential assembly.
Answer B is wrong. The bearing caps should always be marked to the housing to ensure proper installation.
Answer C is wrong. The shim packs and bearing races should be kept in order.
On some types of axles, the pinion gear and differential assembly all come out of the axle housing as one assembly. The bearing caps should always be marked when removed to ensure they go back together correctly.
The axle shafts must be removed before the differential assembly will come out. The shim packs and bearing races should be kept in order.
To get a rotating torque measurement on the pinion gear, the differential case/carrier must be removed.
Measure the rotating torque of the axle assembly if the load of the case, ring gear, and the axle shafts are included. Use an inch-pound torque wrench with a needle-type indicator to get an accurate reading.
Answer D is correct. On some vehicles, the pinion and differential assembly are removed as an assembly.

Question #48
Answer A is wrong. The best way to check for wear is to measure clearances.
Answer B is wrong. The best way to check for wear is to measure clearances.
Answer C is wrong. The best way to check for wear is to measure clearances.
Specification measurements should be taken and recorded to aid in the installation of parts that have tolerances.
All parts should be cleaned and lubricated before the assembly of the part. All components should be inspected for wear or damage.
The annulus gear is locked to the case so that it cannot rotate. In four-wheel drive (4WD) low, the transfer case input shaft is driving the sun gear, which, in turn, is driving the planetary carrier.
Answer D is correct. The best way to check for wear is to measure clearances.

Question #49
Answer A is wrong. Although sealers and gaskets do a good job of sealing the mating surfaces, the surfaces must be clean and free of burrs, or leakage may occur.
Answer B is correct. Only Technician B is correct. Machined transaxle case halves require a sealer or gasket.
Answer C is wrong. Only Technician B is correct.
Answer D is wrong. Only Technician B is correct. Transaxle cases have machined mating surfaces with a very smooth flat finish on them. Not all require a gasket, but they do require some sort of sealant that should be equivalent to the manufacturer's specifications.
If a transaxle is assembled without following manufacturer's sealing instructions, leakage from the case will result.
The transaxle mating surfaces should be inspected for warpage with a straightedge before assembly to ensure a proper fit.

Question #50
Answer A is wrong. Technician B is also correct. Some transfer cases are assembled with paper gaskets.
Answer B is wrong. Technician A is also correct. Most transfer cases are assembled with RTV sealant.
Answer C is correct. Both Technicians are correct.
Answer D is wrong. Both Technicians are correct. Transaxle cases have machined mating surfaces that have a very smooth flat finish on them. Not all require a gasket, but they do require some sort of sealant that should be equivalent to the manufacturer's specifications. If a transaxle is assembled without following manufacturer's sealing instructions, leakage from the case will result.
The transaxle mating surfaces should be inspected for warpage with a straightedge before assembly to ensure a proper fit.

Question #51
Answer A is wrong. Technician B is also correct. A bent input shaft or clutch disc could cause a misalignment between the engine and transmission.
Answer B is wrong. Technician A is also correct. A misaligned engine and transmission would cause the input shaft to put undue stress on the pilot bearing or bushing.
Answer C is correct. Both Technicians are correct.
Answer D is wrong. Both Technicians are correct. The clutch will not make even contact with the flywheel if the bell housing is not aligned properly with the engine block because of something being pinched between them or because of a burr or imperfection on one of the mating surfaces, worn alignment dowels, or loose bell housing bolts. This will cause the clutch to chatter and grab because of the uneven contact when the clutch is being engaged. Reduced clutch pedal free play, growling noises, or vibrations at high speeds are not symptoms of a misalignment condition.
If a clutch disc or an input shaft is bent due to careless removal or installation of the transmission, this will cause misalignment problems. If the transmission is misaligned when it is installed, this can cause the pilot bushing or bearing to wear out prematurely.

Question #52
Answer A is wrong. Starter alignment may cause this problem.
Answer B is wrong. Missing engine dowels may cause this problem.
Answer C is wrong. A damaged or rubbing inspection cover may cause this problem.
The clutch will not make even contact with the flywheel if the bell housing is not aligned properly with the engine block because of something being pinched between them or because of a burr or imperfection on one of the mating surfaces, worn alignment dowels, or loose bell housing bolts. The uneven contact when the clutch is being engaged will cause the clutch to chatter and grab. Reduced clutch pedal free play, growling noises, or vibrations at high speeds are not symptoms of a misalignment condition.
Answer D is correct. A worn transmission main shaft bearing would not cause a grinding noise when trying to start the engine.

Question #53
Answer A is wrong. This will not affect flywheel runout.
Answer B is correct. Crankshaft end play can influence flywheel runout.
Answer C is wrong. This will not affect flywheel runout.
Answer D is wrong. This will not affect flywheel runout.
To measure crankshaft end play, mount the magnetic dial indicator on the back of the engine block. Position the dial indicator to the flywheel. Push the flywheel toward the front of the engine until it stops.
Adjust the dial indicator to zero and then pull the flywheel toward the back of the engine. The reading on the dial indicator will be the crankshaft end play.
The flywheel can be checked with the dial indicator setup after checking the crankshaft end play. Observe the dial indicator as the flywheel is rotated. The measured movement on the dial indicator is the runout; it should be compared to the vehicle specifications.
Excessive main bearing wear will cause low oil pressure or a rear main oil leak, possibly causing the clutch disc to become contaminated with engine oil. A thrust bearing is placed between the crankshaft main bearing cap and the side of a crankshaft journal. These thrust bearings are put into place to control the forward and rearward movement of the crankshaft during acceleration and deceleration. The proper thickness thrust bearing is selected when the engine is assembled to set the crankshaft end play.

Question #54
Answer A is wrong. Third gear may cause this wear.
Answer B is wrong. Fourth gear may cause this wear.
Answer C is wrong. The third and fourth blocking rings may cause this wear.
Worn dogteeth on the third-speed gear or blocking ring may cause hard shifting or jumping out of third gear; however, this problem would not cause a growling noise while driving in third gear. Worn threads in a third-speed blocking ring may cause hard shifting; however, this wear would not cause a growling noise in third gear. Worn, chipped teeth on the third speed gear could result in a growling noise while driving in third gear.
Answer D is correct. The 3/4 synchronizer hub would be the least likely cause of this wear.

Question #55
Answer A is wrong. Technician B is also correct. You can check flywheel runout with this setup.
Answer B is wrong. Technician A is also correct. You can check crankshaft end-play with this setup.
Answer C is correct. Both Technicians are correct.
Answer D is wrong. Both Technicians are correct.
To measure crankshaft end play, mount the magnetic dial indicator on the back of the engine block. Position the dial indicator to the flywheel. Push the flywheel toward the front of the engine until it stops.
Adjust the dial indicator to zero and then pull the flywheel toward the back of the engine. The reading on the dial indicator will be the crankshaft end play.
The flywheel can be checked with the dial indicator setup after checking the crankshaft end play. Observe the dial indicator as the flywheel is rotated. The measured movement on the dial indicator is the runout and should be compared to the vehicle specifications.
Excessive main bearing wear will cause low oil pressure or a rear main oil leak, possibly causing the clutch disc to become contaminated with engine oil. A thrust bearing is placed between the crankshaft main bearing cap and the side of a crankshaft journal. These thrust bearings are put into place to control the forward and rearward movement of the crankshaft during acceleration and deceleration. The proper thickness thrust bearing is selected when the engine is assembled to set the crankshaft end play.

Question #56
Answer A is correct. A straightedge is the best way to check the mating surfaces.
Answer B is wrong. A straightedge is the best way to check the mating surfaces.
Answer C is wrong. A straightedge is the best way to check the mating surfaces.
Answer D is wrong. A straightedge is the best way to check the mating surfaces.
Transaxle cases have machined mating surfaces that have a very smooth flat finish on them. Not all require a gasket, but they do require some sort of sealant that should be equivalent to the manufacturer's specifications. If a transaxle is assembled without following manufacturer's sealing instructions, leakage from the case will result.
The transaxle mating surfaces should be inspected for warpage with a straightedge before assembly to ensure a proper fit.

Question #57
Answer A is wrong. Spider gears are in the differential and would not be affected.
Answer B is wrong. The transmission mount might be the cause, but it will not be damaged.
Answer C is correct. U-joints are Most-Likely to be damaged by excessive driveshaft angle.
Answer D is wrong. Wheel balance will not be affected; so the tires will not be affected.
Driveshaft working angle is also known as pinion angle. The engine and transmission are installed in the chassis at a preset angle usually pointing down at the tailshaft in rear-wheel drive applications. For our purposes, let's assume that the tailshaft is pointing down 3°. To avoid a humming or droning vibration in the driveline, the pinion must be pointing up 3°. This creates a situation where the planes in which they operate are parallel, providing the smoothest U-joint operation. (Note that most manufacturers give these measurement specifications only for pick-up trucks. Unless the manufacturer specifies differently, the static setting of each should be the same.) Causes of driveshaft working angle problems are: sagging (damaged) rear springs, sagging engine or transmission mounts, or major changes in ride height up or down.

Question #58
Answer A is wrong. Technician B is also correct. The runout is checked with a dial indicator.
Answer B is wrong. Technician A is also correct. The axle must be rotated slowly to see the changes in the runout.
Answer C is correct. Both Technicians are correct.
Answer D is wrong. Both Technicians are correct. To measure the axle shaft end play, you will need to remove the wheel and tire assembly and the brake drum. A dial indicator is mounted or clamped to the axle housing or suspension. The axle shaft must be completely pushed into the housing until it stops. Rest the dial indicator head on the face of the axle shaft flange. With the dial indicator set to zero, pull out on the axle shaft. The resulting dial indicator reading is the axle shaft end play. The vehicle differential cover does not need to be removed.
Excessive runout could be caused by a bent axle shaft. A worn C-lock will cause excessive end play. A worn bearing will cause fluid leakage. A bent housing is considered major damage and is noticeable.

Question #59
Answer A is wrong. Torsional dampening springs should face the pressure plate.
Answer B is correct. Only Technician B is correct. Torsional dampening springs help smooth out engine pulsations.
Answer C is wrong. Only Technician B is correct.
Answer D is wrong. Only Technician B is correct. Proper installation of the clutch disc is critical; the damper spring offset must face the transmission. The clutch disc is normally marked to indicate which side should face the flywheel. If the damper spring offset is toward the engine, the springs may contact the flywheel or the flywheel bolts, damaging these components.

Question #60
Answer A is wrong. A shift fork is connected to the synchronizer assembly.
Answer B is wrong. A shift fork is not connected to the counter shaft.
Answer C is wrong. A shift fork is not connected to the blocker rings.
The shift forks are used to shift gears, but they are not connected to the forward gears or reverse gear. Also, the shift forks are not connected to the blocker ring s or the countershaft. A blocker ring is used to stop a gear from spinning while the vehicle is in motion for gear synchronization. A countershaft is used to change the rotation of the gears on the main shaft and for different gear ratios. The shift forks are connected to the synchronizer sleeve. They move the synchronizer sleeve forward or backward to engage the transaxle in a gear.
Answer D is correct. A shift fork is connected to the synchronizer sleeve.

Question #61
Answer A is wrong. 0.005 inch (0.127 mm) is less than minimum.
Answer B is wrong. 0.008 inch (0.203 mm) is less than minimum.
Answer C is correct. 0.012 inch (0.305 mm) is the minimum.
Answer D is wrong. 0.025 inch (0.638 mm) is more than minimum.
If the clutch facing is worn too thin, the clamping force of the pressure plate will not be as much as it was when the clutch disc was at full thickness. There will not be enough spring pressure left in the pressure plate to maintain a hard clamping force on the clutch disc because the springs in the pressure plate are fully extended and are not applying enough pressure. The minimum thickness of the clutch disc lining is 0.012 inch (0.3 mm). Slippage will occur if the lining is any thinner.

Question #62

Answer A is correct. Only Technician A is correct. The pressure plate should be replaced, not resurfaced.

Answer B is wrong. The pressure plate should be replaced, not resurfaced.

Answer C is wrong. Only Technician A is correct.

Answer D is wrong. Only Technician A is correct. If too much material is removed from the flywheel, the torsion springs on the clutch plate are moved closer to the mounting bolts on the flywheel; these springs may contact the heads of the flywheel bolts.

Removing excessive material from the flywheel moves the pressure plate forward, away from the release bearing. This action increases free play so the slave cylinder rod may not move far enough to release the clutch. By not resurfacing the flywheel, damage or premature wear can be caused to the rest of the new assembly that was installed. A flywheel should be resurfaced every time a clutch assembly is replaced.

Resurfacing the flywheel will ensure that it has the flatness and a microfinish it needs so that the new clutch disc breaks in properly. If it is not resurfaced, the clutch disc will probably glaze and chatter. The flywheel should be cleaned with hot water and soap to remove all residue on the surface after it has been resurfaced.

Question #63

Answer A is correct. Only Technician A is correct. The bent shift fork should be discarded and a new one installed.

Answer B is wrong. The bent shift fork should be discarded and a new one installed.

Answer C is wrong. Only Technician A is correct.

Answer D is wrong. Only Technician A is correct. The shift forks are used to shift gears, but they are not connected to the forward gears or reverse gear. Also, the shift forks are not connected to the blocker rings or the countershaft. A blocker ring is used to stop a gear from spinning while the vehicle is in motion for gear synchronization. A countershaft is used to change the rotation of the gears on the main shaft and for different gear ratios. The shift forks are connected to the synchronizer sleeve. They move the synchronizer sleeve forward or backward to engage the transaxle in a gear.

Question #64

Answer A is wrong. The ring gear tooth pattern has nothing to do with carrier bearing preload.

Answer B is correct. Only Technician B is correct. The ring gear tooth pattern also checks pinion depth.

Answer C is wrong. Only Technician B is correct.

Answer D is wrong. Only Technician B is correct. Prussian blue (a marking compound) is commonly used to check the tooth contact pattern on gear setup.

A used gear set will have a shiny pattern on the gear teeth that can be visually inspected. Look for the pattern to be centered with either new or used gears after assembly. Carrier-bearing adjustment may be necessary to correct the depth and backlash of the gears. Do not concern yourself with the exact appearance of the pattern, as it varies by gear vendor. Take some time to look at the location of the pattern on the ring and pinion faces for your favorite manufacturer.

Question #65

Answer A is wrong. The flywheel has no fixed relationship to the transmission input shaft.

Answer B is correct. The flywheel has an important relationship to the crankshaft because it is a major factor in engine balance.

Answer C is wrong. The flywheel has no fixed relationship to the clutch disc.

Answer D is wrong. The flywheel has no fixed relationship to the bell housing. If the flywheel is scored, burned, worn, or has heat cracks, it should be resurfaced or replaced. Heat cracks almost always require flywheel replacement. During resurfacing of a flywheel, 0.0010 -0.040 inch (.025mm-1mm) of material can be safely removed from the flywheel surface. Do not remove excessive material from the flywheel surface.

Worn or chipped flywheel ring gear teeth may cause improper flywheel balance and engine vibrations.

Starter motor damage and engagement problems can also result from worn, chipped, or missing flywheel teeth. When the flywheel is removed from the crankshaft, punch index marks on the flywheel and crankshaft so that the flywheel can be reinstalled in the original position. The will maintain proper balance and reduce the chances of a vibration being induced into the flywheel/clutch system.

Question #66
Answer A is wrong. The ring gear does not contact the clutch.
Answer B is correct. Only Technician B is correct. The ring gear can affect the starter engagement.
Answer C is wrong. Only Technician B is correct.
Answer D is wrong. Only Technician B is correct. If the flywheel is scored, burned, worn or has heat cracks, it should be resurfaced or replaced. Heat cracks almost always require flywheel replacement. During resurfacing of a flywheel, 0.0010 -0.040 inch (0.025mm-1mm) of material can be safely removed from the flywheel surface. Do not remove excessive material from the flywheel surface.
Worn or chipped flywheel ring gear teeth may cause improper flywheel balance and engine vibrations.
Starter motor damage and engagement problems can also result from worn, chipped, or missing flywheel teeth. When the flywheel is removed from the crankshaft, punch index marks on the flywheel and crankshaft so that the flywheel can be reinstalled in the original position. The will maintain proper balance and reduce the chances of a vibration being induced into the flywheel/clutch system.

Question #67
Answer A is wrong. Technician B is also correct. These gears are shimmed for proper fit and clearance. The best way to check for wear is to measure clearances.
Answer B is wrong. Technician A is also correct. These gears turn when the vehicle turns.
Answer C is correct. Both Technicians are correct.
Answer D is wrong. Both Technicians are correct. If there were a constant whining noise coming from the differential, the noise could not be coming from the side gears or the spider gears. These gears are used only when the vehicle is turning; so if they were damaged, the noise would only be heard on turns. The wrong differential lube could cause damage to the differential parts, but will not cause a whining noise.

Question #68
Answer A is wrong. Technician B is also correct. The differential gears turn only when the vehicle is moving.
Answer B is wrong. Technician A is also correct. Some differential noises can be mistaken for other vehicle noises.
Answer C is correct. Both Technicians are correct.
Answer D is wrong. Both Technicians are correct. Damaged ring gear teeth would cause a clicking noise while the vehicle is in motion. This problem would not cause differential chatter. Improper preload on differential components, such as side bearings, may cause differential chatter. If there were a constant whining noise coming from the differential, the noise could not be coming from the side gears or the spider gears. These gears are used only when the vehicle is turning, so if they were damaged, the noise would be heard only on turns. The wrong differential lube could cause damage to the differential parts, but it will not cause a whining noise. If the preload and backlash are not set properly, the gear mesh could be too tight and cause a whining noise.

Question #69
Answer A is correct. Only Technician A is correct. A damaged cluster/countershaft gear could cause a clunking noise in first and Reverse.
Answer B is wrong. The reverse idler gear does not make contact in first; therefore, it would not make any noise in first.
Answer C is wrong. Only Technician A is correct.
Answer D is wrong. Only Technician A is correct. The inspection performed on the input shaft and gears should be repeated on the output shaft and gears.
In some transaxles, the output shaft assembly is serviced as a complete unit. If any gear or component on the output shaft is worn, the complete assembly must be replaced. Use a feeler gauge to measure the clearance between the dogteeth on the first- and second-speed gear and the matching blocking ring. Inspect the output shaft bearings in the transaxle case for roughness or looseness. Worn, chipped teeth on the third-speed gear could result in a growling noise while driving in third gear. Inspect the reverse idler gear teeth for chips, pits, and cracks. Although worn reverse idler teeth may cause a growling noise while driving in reverse, this problem would not cause a failure to shift into reverse.

Question #70
Answer A is wrong. The clutch pedal will not have any free play in a self-adjusting system.
Answer B is wrong. The clutch pedal will not have any free play in a self-adjusting system.
Answer C is correct. Self-adjusting clutches do not normally have any free play, which causes the release bearing to contact the clutch release mechanism continuously.
Answer D is wrong. A clutch with a self-adjusting cable does not have an over-center assist spring. Some vehicles have linkages connected from the clutch pedal to the release fork. In other clutch systems, the clutch pedal is connected to the release fork by a cable. Another popular method designed to operate the clutch is a hydraulically controlled system. Many clutch linkages or cables have an adjustment to set the clutch pedal free play. Clutch pedal free play is the amount of pedal movement before the release bearing contacts the pressure plate release fingers or diaphragm. Many late-model vehicles have a self-adjusting clutch cable. In these clutches the cable is wrapped around and attached to a toothed wheel, and a ratcheting spring-loaded pawl is engaged with the toothed wheel. Each time the clutch pedal is released, the pawl removes any slack from the cable by engaging the next tooth on the wheel. The self-adjusting clutch cable system and the hydraulic clutch system have no built-in free play, and they use a constant running release bearing. The clutch release bearing is connected to the release fork, and the center opening in this bearing is mounted on a machined hub that is bolted to the front of the transmission.

Question #71
Answer A is wrong. Technician B is also correct. Improper adjustment may not disengage the clutch, causing the gears to grind.
Answer B is wrong. Technician A is also correct. Improper adjustment may not let the clutch fully engage.
Answer C is correct. Both Technicians are correct.
Answer D is wrong. Both Technicians A and B are correct. When there is no clutch pedal free play, the clutch is not fully engaged. The release bearing is touching the fingers of the pressure plate. This will relieve some of the clamping on the clutch disc. This may cause the clutch to slip; it will not cause hard shifting, improper clutch release, or transaxle gear damage. These would all be signs of too much clutch pedal free play.
Many late-model vehicles have self-adjusting cables. The cable is adjusted when the clutch pedal is released as the clutch disc wears from normal use. These systems use a constant-running release bearing.
It is always in contact with the pressure plate. The clutch pedal will not have any clutch pedal free play.

Question #72
Answer A is wrong. A worn second gear may cause the transaxle to jump out of second gear.
Answer B is wrong. Excessive main shaft end play will cause the transaxle to jump out of second gear.
Answer C is wrong. A shifter linkage out of adjustment may cause the transaxle to jump out of second gear.
Worn dogteeth on the second-speed gear and blocking ring would not result in a growling noise while driving, or accelerating, in second gear, but this problem could not cause hard shifting in second gear. Worn dogteeth on the second-speed gear and blocking ring may cause the transaxle to jump out of second gear.
Inspect all small parts in the transmission for wear. A service manual may list specifications for the thickness of parts, such as thrust washers. If specifications are not available, inspect each part for signs of wear or breakage. Normally all the snap-rings, roller bearings, washers, and spacers are replaced during a transaxle overhaul. Most manufacturers sell a small parts kit that includes all of these parts.
Answer D is correct. A worn ring gear in the differential will not cause the transaxle to jump out of second gear.

Question #73
Answer A is correct. Only Technician A is correct. Snap-rings and spacers are usually replaced during rebuilding.
Answer B is wrong. Snap-rings and spacers can be obtained in a small kit.
Answer C is wrong. Only Technician A is correct.
Answer D is wrong. Only Technician A is correct. Inspect all small parts in the transmission for wear. A service manual may list specifications for the thickness of parts, such as thrust washers. If specifications are not available, inspect each part for signs of wear or breakage. Normally all the snap-rings, roller bearings, washers, and spacers are replaced during a transaxle overhaul. Most manufacturers sell a small parts kit that includes all of these parts.

Question #74
Answer A is wrong. A worn bearing could cause a growling noise.
Answer B is wrong. A worn bearing could cause a growling noise.
Answer C is wrong. A worn bearing could cause a growling noise.
Noises in differentials can be from the axle bearings, ring and pinion, any of the differential bearings, or differential side gears. Axle bearings usually can be isolated by the change in noise as the vehicle experiences different side loads or by raising the vehicle and running it in gear. Most axle bearing noises will subside dramatically when weight is taken off the wheels. Ring and pinion noises are associated with a whine or growl that changes in pitch as vehicle speed or engine load changes. To diagnose differential carrier and pinion bearing noises, a stethoscope is often employed to pinpoint the location of the noise. Since the side gears are turning only while cornering, they do not cause a whining noise while driving straight ahead.
Answer D is correct. Worn spider gears are the LEAST-Likely to cause growling on deceleration.

Question #75
Answer A is correct. A thicker shim will increase the turning torque.
Answer B is wrong. A thinner shim should be used if the turning torque is more than specified.
Answer C is wrong. A thinner shim should be used if the turning torque is more than specified.
Answer D is wrong. The thicker shim should be installed only in the bell housing side of the transfer case.
While measuring the differential side play to determine the required side bearing shim thickness, a new bearing cup is installed in the transaxle case without the shim. The proper shim thickness is equal to the differential end play plus a specified thickness for bearing preload. While measuring the differential end play, the transaxle case bolts must be tightened to the specified torque. A medium load should be applied to the differential in the upward direction while measuring the differential end play.

Question #76
Answer A is wrong. The center support bearing will not make noise if the vehicle is stopped.
Answer B is correct. Only Technician B is correct. Some center support bearings need to be lubricated.
Answer C is wrong. Only Technician B is correct.
Answer D is wrong. Only Technician B is correct. A worn driveshaft center bearing or an outer rear axle bearing causes a growling noise that is not influenced by acceleration and deceleration.
A center support bearing is usually maintenance-free, and it is usually seated. A center support bearing is not part of the driveshaft, but is part of the driveline. A center support bearing is found mainly on trucks and vans and is used to shorten the length of long driveshafts and to decrease pinion angles.

Question #77
Answer A is wrong. Technician B is also correct. Bushings are used in some reverse idler gears in manual transmissions/transaxles.
Answer B is wrong. Technician A is also correct. Needle bearings are used in most transmissions/transaxles.
Answer C is correct. Both Technicians are correct.
Answer D is wrong. Both Technicians are correct. Inspect the reverse idler gear teeth for chips, pits, and cracks. Although worn reverse idler teeth may cause a growling noise while driving in reverse, this problem would not cause a failure to shift into reverse. A broken reverse shifter fork may not allow the transaxle to shift into reverse without causing noise.
The needle bearings on a reverse idler gear should be smooth and shiny. Carefully inspect the needle bearings for wear, burrs, and other problems. Worn or damaged needle bearings should be replaced, or damage to other components may occur.

Question #78
Answer A is wrong. Technician B is also correct. These gears turn when the vehicle moves.
Answer B is wrong. Technician A is also correct. Some component that has contact with the shaft must have worn the shaft. If the shaft is worn, the case and all related components must be checked.
Answer C is correct. Both Technicians are correct.
Answer D is wrong. Both Technicians are correct. Any transaxle needle bearings for a gear should be smooth and shiny. Carefully inspect the needle bearings for wear, burrs, and other problems. Worn or damaged needle bearings should be replaced, or damage to other components may occur. Shafts that the needle bearings ride on should also be carefully inspected for scoring, wear, pitting, and other imperfections. Shafts and bearings should be replaced as matched sets.

Question #79
Answer A is wrong. Normally when removing the driveshaft, two marks need to be applied; one on the driveshaft, and the other on the pinion flange.
Answer B is wrong. When replacing the U-joints, the driveshaft does not need to be marked.
Answer C is wrong. Driveshaft runout measurement is performed with a dial indicator.
Prior to driveshaft balancing always inspect this shaft for damage, a missing balance weight, or an accumulation of dirt or undercoating. Remove both rear wheel assemblies and install the wheel nuts on the wheel studs with the flat side toward the brake drums. Chalk mark the driveshaft at four locations 90 degrees apart, just forward of the driveshaft balance weight. Fabricate a tool to hold the strobe light pickup against the rear axle housing just behind the pinion yoke. Run the vehicle in gear until the driveshaft vibration is most severe. Do not exceed 55 mph, because 55 mph indicated on the speedometer has a possible tire speed of 110 mph. Point the strobe light at the chalk marks on the driveshaft and note the position of one reference mark.
Answer D is correct. Driveshaft balance is checked with four marks 90 degrees apart.

Question #80
Answer A is wrong. Technician B is also correct. A different gear may have a different number of teeth, which would change the speedometer reading.
Answer B is wrong. Technician A is also correct. When replacing a speedometer cable core, the old gear can be used.
Answer C is correct. Both Technicians are correct.
Answer D is wrong. Both Technicians are correct. A speedometer gear is normally mounted on the output shaft. The output shaft spins at driveline speed. An output shaft will never have a drive gear machined into the output shaft. A drive gear is made out of a plastic nylon-type gear, and the teeth on the drive and driven gear have a helical-type cut on them. The helical-type cut and the plastic-type gear are used for quiet operation.
Whenever a speedometer drive gear is replaced, the driven gear also should be replaced. If a speedometer cable assembly core is damaged, it may be replaced with a new core. The new core must be cut to the same size as the one being replaced and must be properly lubricated before installation into the cable housing.

Question #81
Answer A is correct. The engine will need to be supported without the transaxle.
Answer B is wrong. The engine oil does not need to be drained when removing the transaxle.
Answer C is wrong. The negative battery cable should be disconnected; not the positive.
Answer D is wrong. The engine does not need to be removed when removing the transaxle from many vehicles.
Prior to transaxle removal, the battery ground cable, shift linkages or cables, speedometer cable, vehicle speed sensor, and all electrical connections must be disconnected. Drain the transaxle lubricant. On front wheel drive vehicles, the front drive axle must be removed from the transaxle prior to transaxle removal.
Before the transaxle retaining bolts are loosened, an engine support fixture must be installed to support the weight of the engine. On some front-wheel-drive vehicles the engine cradle, or part of the cradle, must be removed prior to transaxle removal. Use a transmission jack to support the weight of the transaxle during the removal process.

Question #82
Answer A is wrong. The axle shafts must be removed from the transaxle to remove the transaxle differential.
Answer B is wrong. The transaxle must be removed from the vehicle and the case opened to remove the transaxle differential.
Answer C is correct. The engine does not have to be removed.
Answer D is wrong. The lower control arms may need to be removed.
Remove the retaining bolts between the two transaxle halves, and separate these halves. Be careful not to mar the transaxle case. Remove the differential assembly from the case. Remove the reverse idler shaft bolt, reverse idler gear shaft, and reverse fork bracket. Remove the selector shaft spacer and selector shaft. Remove the transaxle cover and use a pair of nap ring pliers to remove the snap-rings in the input and output shafts. Support the transaxle on the proper bench fixture and shims. Use the proper driving fixture to press the input and output shaft assemblies out of the case. Remove the reverse brake shim, friction cone, blocking ring, and needle bearing. Wash all of the transaxle components in an approved cleaning solution and blow-dry these components with compressed air.

Question #83
Answer A is wrong. The gear is mounted on the output shaft and is not to be mistaken for the driven gear that is part of the vehicle speed sensor or speedometer cable.
Answer B is wrong. The drive gear is made out of nylon.
Answer C is correct. The speedometer gear is NOT machined into the output shaft. The drive gear is made of nylon, is mounted on the output shaft, and does have helical teeth.
Answer D is wrong. The gear has helical teeth.
A speedometer gear is normally mounted on the output shaft. The output shaft spins at driveline speed. An output shaft will never have a drive gear machined into the output shaft. A drive gear is made out of a plastic nylon-type gear, and the teeth on the drive and driven gear have a helical-type cut on them. The helical-type cut and the plastic-type gear are used for quiet operation.
Whenever a speedometer drive gear is replaced, the driven gear also should be replaced. If a speedometer cable assembly core is damaged, it may be replaced with a new core. The new core must be cut to the same size as the one being replaced, and it must be properly lubricated before installation into the cable housing.

Question #84
Answer A is correct. The transaxle must be removed.
Answer B is wrong. The transaxle case should be split after it is removed from the vehicle.
Answer C is wrong. The engine should not need to be removed to remove the transaxle.
Answer D is wrong. Only part of the transaxle needs to be disassembled to remove the differential assembly.
Remove the retaining bolts between the two transaxle halves, and separate these halves. Be careful not to mar the transaxle case. Remove the differential assembly from the case. Remove the reverse idler shaft bolt, reverse idler gear shaft, and reverse fork bracket. Remove the selector shaft spacer and selector shaft.
Remove the transaxle cover and use a pair of nap ring pliers to remove the snap-rings in the input and output shafts. Support the transaxle on the proper bench fixture and shims. Use the proper driving fixture to press the input and output shaft assemblies out of the case. Remove the reverse brake shim, friction cone, blocking ring, and needle bearing. Wash all of the transaxle components in an approved cleaning solution and blow-dry these components with compressed air.

Question #85

Answer A is wrong. Stripped drive and driven gears could cause an inoperative speedometer.

Answer B is wrong. If the drive gear slips on the end of the sensor, the speedometer may be inoperative.

Answer C is correct. The drive gear position is not adjustable.

Answer D is wrong. If the driven gear teeth are stripped, the speedometer will be inoperative.

A speedometer gear is normally mounted on the output shaft. The output shaft spins at driveline speed.

An output shaft will never have a drive gear machined into the output shaft. A drive gear is made out of a plastic nylon-type gear, and the teeth on the drive and driven gear have a helical-type cut on them. The helical-type cut and the plastic-type gear are used for quiet operation.

Whenever a speedometer drive gear is replaced, the driven gear also should be replaced. If a speedometer cable assembly core is damaged, it may be replaced with a new core. The new core must be cut to the same size as the one being replaced, and it must be properly lubricated before installation into the cable housing.

Question #86

Answer A is wrong. The bearings do not need to be replaced to correct the pinion gear rotating torque.

Answer B is wrong. The pinion gear cannot be replaced without replacing the ring gear.

Answer C is wrong. The pinion gear nut should never be loosened to compensate for incorrect pinion gear rotating torque.

Coat the outside diameter of the new seal with gasket sealer, and lubricate the seal lips and bearings with the manufacturer's specified differential lubricant. Use the proper seal driver to install the pinion seal.

Install the differential flange and a new pinion nut. Since the pinion nut torque determines the pinion bearing preload, the torque on this nut is critical. Tighten the pinion nut until the manufacturer's specified turning torque is obtained with an inch-pound torque wrench and socket installed on the pinion nut.

Answer D is correct. A new collapsible spacer and nut must be used anytime the turning torque must be reduced.

Question #87

Answer A is wrong. Technician B is also correct. A sign of a bad outer CV joint is a clunking or clicking while turning.

Answer B is wrong. Technician A is also correct. A bad CV joint could also make noise while the vehicle is going straight, if it is excessively worn.

Answer C is correct. Both Technicians are correct.

Answer D is wrong. Both Technicians are correct. Most CV joints will produce a clicking noise only heard during turns until they are completely worn out.

Then they may also make clicking noises while driving straight ahead. They are usually replaced as a complete assembly. Prior to removing the CV joint boot, mark the inner end of the boot in relation to the drive axle so the boot may be installed in the original position. When reassembling the joint always install the grease in the joint that is provided in the repair kit.

Question #88

Answer A is wrong. The pinion seal should always be replaced with the pinion gear.

Answer B is wrong. The collapsible spacer is not reusable.

Answer C is wrong. It is always wise to replace the axle seals whenever the axles are removed.

If you are not familiar with the process of rebuilding differential, it would be wise to read through any manufacturer's procedure, as it exceeds the scope of this book. Key issues that you should know are: The ring and pinion set must be replaced as a set; collapsible spacers must be replaced if they are over-tightened or when the pinion bearings need to be replaced; and bearings should be replaced as assemblies.

When determining the condition of a ring and pinion, you are looking for excessive wear on the gears, pitting or grooves worn in the gear faces, or evidence that the gears have overheated, which will cause them to turn blue or black.

Answer D is correct. The spider gears are only replaced if they are worn.

Question #89
Answer A is wrong. A plugged axle housing vent may cause excessive pressure that will lead to a leak.
Answer B is wrong. If the differential is overfilled, the fluid may leak past a housing seal.
Answer C is wrong. Worn axle shaft bearings will put a load on the axle shaft seal, and they could cause a leak.
If the differential fluid is too full, excessive pressure may build up and cause the differential fluid to leak past a seal. If the vent becomes plugged, it will cause excessive pressure in the differential housing, and a leak will occur. Axle shaft bearings that are worn will cause the axle shaft to apply load on the axle shaft seal, and the seal will fail. Pinion seals and carrier covers are some other common and easy-to-spot leak points.
Answer D is correct. Fluid type will not normally cause a leak.

Question #90
Answer A is wrong. The universal joint should not be greased so much that grease squirts out of the caps.
Answer B is wrong. Driveshaft balance is not affected by the amount of grease in the universal joint.
Answer C is wrong. Neither Technician is correct. A universal joint has prelube on the inside, but this prelube is not sufficient to lubricate the joint when it has been installed in a vehicle. You should grease a new universal joint when it is installed. When a universal joint is greased, you should not pump so much grease into the joint that it squirts out of the caps. When this happens, it damages the seals around the caps and shortens the life of the joint.
Answer D is correct. Neither Technician is correct.

Question #91
Answer A is wrong. Technician B is also incorrect. An open in the shift switch circuit would not power or signal the transfer case motor to shift the unit into 4WD.
Answer B is wrong. Technician A is also correct. The shift motor drives the shift linkages that are housed inside the transfer case.
Answer C is correct. Both Technicians are correct.
Answer D is wrong. Both Technicians are correct. Electronically controlled actuation and engagements systems require special diagnostic test equipment. As a technician, you should have an understanding of electrical and electronic operational theory. Be sure to refer to manufacturer's recommended testing and service procedures. Damage to electronic components and control devices can result from improper testing and installation procedures.

Question #92
Answer A is wrong. Scored side bearings would give inaccurate readings while measuring case/carrier runout.
Answer B is correct. Only Technician B is correct. Case/carrier runout is part of total ring gear runout.
Answer C is wrong. Only Technician B is correct.
Answer D is wrong. Only Technician B is correct. The ring gear runout should be measured prior to disassembling the differential. Mount the dial indicator assembly, and position the dial indicator stem at a 90-degree angle against the back of the ring gear. Turn the dial indicator to zero and rotate the ring gear one revolution. The difference between the highest and lowest dial indicator reading is the ring gear runout.
To determine if excessive ring rear runout is caused by the ring gear on the case/carrier, remove the ring gear and case/carrier and remove the ring gear from the case/carrier. Install the case/carrier assembly without the ring gear, and be sure the side bearings are in good condition and properly torqued. Position the dial indicator against the backside of the case/carrier, and rotate the case/carrier one revolution to measure the case/carrier runout. If the case/carrier runout is normal but the ring gear runout is excessive, replace the ring gear and pinion. When the case/carrier runout is excessive, replace the case/carrier.

Question #93

Answer A is wrong. The driveshaft must be removed to check pinion flange runout.

Answer B is wrong. A dial indicator would be used to check pinion flange runout.

Answer C is correct. A dial indicator would be used to check pinion flange runout.

Answer D is wrong. Pinion flange runout cannot be checked with the flange removed.

To check pinion flange runout, remove the driveshaft and mount a dial indicator against the face of the flange. Rotate the flange and note the readings on the dial indicator; these are the runout readings. If the flange is removed, you cannot measure pinion flange runout. A loose pinion nut allows pinion shaft end play, resulting in a clunking noise on acceleration and deceleration. Insufficient pinion nut torque will affect the pinion bearing. Preload will not cause a growling noise.

Question #94

Answer A is wrong. A loose pinion nut can affect ring-and-pinion contact and produce a clunk.

Answer B is wrong. A loose pinion nut can increase pinion end play in the bearings and accelerate wear.

Answer C is wrong. A loose pinion nut can affect ring-and-pinion contact and produce the wrong wear pattern.

To check pinion flange runout, remove the driveshaft and mount a dial indicator against the face of the flange. Rotate the flange and note the readings on the dial indicator; these are the runout readings. If the flange is removed, you cannot measure pinion flange runout. A loose pinion nut allows pinion shaft end play, resulting in a clunking noise on acceleration and deceleration. Insufficient pinion nut torque will affect the pinion bearing. Preload will not cause a growling noise.

Answer D is correct. Spider gears have no relation to the tightness of the pinion nut.

Question #95

Answer A is correct. Only Technician A is correct. The spider gears ride on the pinion shaft and should have smooth and shinny bores.

Answer B is wrong. The spider gears do not contain needle bearings.

Answer C is wrong. Only Technician A is correct.

Answer D is wrong. Only Technician A is correct. The side gear end play must be measured individually on each side gear with the thrust washers removed.

Side gears with the specified thrust washer have slight end play, but no preload.

The spider gears ride on the pinion shaft, and the bore in the gears should be smooth, shiny, and have no signs of pits or scuffing. The pinion shaft should also be free of pitting and scoring. There is no needle bearing in any of the spider gears.

Question #96

Answer A is correct. Hunting-type gear sets do not require the gear set to be timed.

Answer B is wrong. Loose ring gear bolts may cause a gear chuckle or knocking while driving.

Answer C is wrong. Damaged ring gear and pinion gear teeth may cause a ticking noise while driving.

Answer D is wrong. On some ring gear sets, the painted teeth on the ring gear must be matched to the painted teeth on the pinion gear.

Many ring gear and pinion gear sets have timing marks that must be aligned when assembling the differential. On some gear sets one pinion gear tooth is grooved and painted, and the ring gear has a notch between two painted teeth. When the ring and pinion gears are meshed together, the grooved and painted pinion gear tooth must fit between the notched and painted ring gear teeth. Some ring gear and pinion gear sets do not have timing marks. These gear sets are referred to as hunting gears.

Question #97

Answer A is wrong. Technician B is also correct. A bad axle bearing would allow the axle to apply pressure to the seal.

Answer B is wrong. Technician A is also correct. The axle seals are not designed to hold back large quantities of axle fluid.

Answer C is correct. Both Technicians are correct.

Answer D is wrong. Both Technicians are correct. If the differential fluid is too full, excessive pressure may build up and cause the differential fluid to leak past a seal. If the vent becomes plugged, it will cause excessive pressure in the differential housing and a leak will occur. Axle shaft bearings that are worn will cause the axle shaft to apply load on the axle shaft seal, and the seal will fail. Pinion seals and carrier covers are some other common and easy-to-spot leak points.

Question #98
Answer A is wrong. A worn blocking ring could cause the fluid to contain gold-colored material.
Answer B is correct. A worn second gear would produce metal shavings in the fluid.
Answer C is wrong. A worn thrust washer could cause the fluid to contain gold-colored material.
Answer D is wrong. A worn shift fork could cause the fluid to contain gold-colored material.
Drain and inspect the fluid during transmission or transaxle overhaul. Gold-color particles in the fluid are produced by the wearing of the brass blocking rings on the synchronizers. Metal shavings in the fluid are produced from the wearing of the gears.
An excessive amount of shavings in the fluid indicates severe gear and synchronizer wear.

Question #99
Answer A is wrong. A rusted shift linkage would cause hard shifting at all times.
Answer B is wrong. If the clutch disc were sticking to the flywheel, the hard shifting would occur at all times.
Answer C is correct. Temperature affects fluid viscosity and transmission shifting when cold.
Answer D is wrong. A bad slave cylinder would cause hard shifting at all times.
Some vehicle manufacturers recommend a mineral oil with an extreme pressure (EP) additive for manual transmissions. The most common gear oil classifications are SAE 75W, 75W-80, 80W-90, 85W-90, 90, or 140. Thicker gear oils have higher classification numbers. Some manual transmissions require engine oil or automatic transmission fluid. Many manual transaxles require 30W engine oil, 90W gear oil or ATF. Using the wrong fluid can cause shifting and/or wear concerns. On a cold start, if the fluid is too thick, the vehicle may be hard to shift. This will also cause the transaxle to get poor lubrication on a cold start.

Question #100
Answer A is wrong. A dial indicator is used to check ring gear runout.
Answer B is wrong. A dial indicator is used to check ring gear runout.
Answer C is wrong. A dial indicator is used to check ring gear runout.
Ring gear runout is best measured before differential disassembly. In most applications a dial indicator is mounted to the carrier or axle housing and the measuring tip is set on the back of the ring gear (the side without gear teeth). Measure runout by turning the ring gear and watching for variations in the dial indicator.
Use manufacturer specifications to determine if the gear has excessive runout. Possible causes for ring gear distortion include: uneven torque of ring gear bolts, overheated gear assembly (usually evident long before you take this measurement), dropped gear during assembly, manufacturing defects, and debris between the carrier and the gear. Torque and debris issues can usually be corrected without replacing the gear, but all other issues require ring and pinion set replacement.
Answer D is correct. A dial indicator is used to check ring gear runout.

Question #101
Answer A is wrong. Locking hubs must be removed for access to the front axles.
Answer B is wrong. If the vehicle has drum brakes on the front wheels, the drums must be removed for access to the front axles.
Answer C is correct. Brake pads or shoes do not need to be removed to check axle end play.
Answer D is wrong. Wheels must be removed for access to the axles.
To measure the axle shaft end play, you will need to remove the wheel and tire assembly and the brake drum. A dial indicator is mounted or clamped to the axle housing or suspension. The axle shaft must be completely pushed in to the housing until it stops. Rest the dial indicator head on the face of the axle shaft flange. With the dial indicator set to zero, pull out on the axle shaft. The resulting dial indicator reading is the axle shaft end play. The vehicle differential cover does not need to be removed.
Excessive runout could be caused by a bent axle shaft. A worn C-lock will cause excessive end play. A worn bearing will cause fluid leakage. A bent housing is considered major damage and is noticeable.

Question #102

Answer A is wrong. The torque of the whole assembly would be measured if the measurement were taken with the axle fully assembled.

Answer B is wrong. The axle shafts do not need to be removed to measure pinion bearing preload.

Answer C is correct. Only the pinion bearing preload torque is wanted.

Answer D is wrong. An inch-pound torque wrench with a needle-type indicator should be used to measure pinion bearing preload.

Coat the outside diameter of the new seal with gasket sealer, and lubricate the seal lips and bearings with the manufacturer's specified differential lubricant. Use the proper seal driver to install the pinion seal.

Install the differential flange and a new pinion nut. Since the pinion nut torque determines the pinion bearing preload, the torque on this nut is critical. The differential case/carrier is not installed at this point.

Tighten the pinion nut until the manufacturer's specified turning torque is obtained with an inch-pound torque wrench and socket installed on the pinion nut.

Question #103

Answer A is wrong. In any position other than choice D, centrifugal force will try to throw grease out of the fitting.

Answer B is wrong. In any position other than choice D, centrifugal force will try to throw grease out of the fitting.

Answer C is wrong. In any position other than choice D, centrifugal force will try to throw grease out of the fitting.

Prior to driveshaft removal, always mark the driveshaft in relation to the differential flange so this shaft may be installed in the original position. After the universal join retaining clips are removed from the driveshaft, a vice and the proper size of socket may be used to remove the spider from the yoke.

A universal joint has prelube on the inside, but this prelube is not sufficient to lubricate the joint when it has been installed in a vehicle. You should grease a new universal joint when it is installed. When a universal joint is greased, you should not pump so much grease into the joint that it squirts out of the caps. When this happens, it damages the seals around the caps and shortens the life of the joint.

Answer D is correct. This is the only position in which centrifugal force will not try to throw grease out of the fitting.

Question #104

Answer A is wrong. The CV joint will be open and cleaned to replace the boot and grease.

Answer B is correct. The wheel bearing is LEAST-Likely to be inspected when replacing a CV boot.

Answer C is wrong. The CV joint attaches to the drive axle shaft.

Answer D is wrong. When the axle is out, the axle seals should be checked. If they are hard or show any damage, they should be replaced while the customer is paying the bill, rather than a few days later when they develop a leak.

Most CV joints are replaced as a complete assembly. Prior to removing the CV joint boot, mark the inner end of the boot in relation to the drive axle so that the boot may be installed in the original position.

When reassembling the joint, always install the grease in the joint that is provided in the repair kit. An outer CV joint will make a clicking noise when the vehicle is turning. This means the joint ball bearings are bad and the grease is contaminated.

A worn front wheel bearing usually results in a growling noise while cornering or driving straight ahead.

Question #105
Answer A is wrong. The differential side bearings are pressed on to the differential and therefore held in place by an interference fit.
Answer B is wrong. The differential side bearings are pressed on to the differential and therefore held in place by an interference fit.
Answer C is correct. The differential side bearings are pressed on to the differential and therefore held in place by an interference fit.
Answer D is wrong. The differential side bearings are pressed on to the differential and therefore held in place by an interference fit.
A side bearing preload adjustment shim is positioned behind one of the side bearing cups. When the differential side bearings are being replaced, they do not need to be packed with grease. The bearings should be lubricated with the differential lube. The differential case/carrier does not need to be replaced when the bearings are replaced. The only time a case is replaced is when it shows signs of damage. When new bearings are installed they should be installed using a hydraulic press, and the bearing races should be replaced also. The new bearing will not wear in properly if the old race is used.

Question #106
Answer A is wrong. Differential side bearings do not need to be packed with grease.
Answer B is wrong. A hydraulic press should be used to install the differential side bearings.
Answer C is correct. If the bearing race is reused, premature bearing failure may occur.
Answer D is wrong. The case should be replaced only if there are signs of damage.
A side-bearing preload adjustment shim is positioned behind one of the side-bearing cups. When the differential side bearings are being replaced, they do not need to be packed with grease. The bearings should be lubricated with the differential lube. The differential case/carrier does not need to be replaced when the bearings are replaced. The only time a case is replaced is when it shows signs of damage. When new bearings are installed, they should be installed using a hydraulic press, and the bearing races should be replaced also. The new bearing will not wear in properly if the old race is used.

Question #107
Answer A is wrong. When the left-side adjuster nut is loosened, the backlash is increased.
Answer B is correct. Only Technician B is correct. Pinion preload on some differentials is adjusted with selectable shims.
Answer C is wrong. Technician A is wrong.
Answer D is wrong. Only Technician B is correct. When the differential has threaded adjusters on the outside of the side bearings, loosen the right adjuster and tighten the left adjuster to obtain zero backlash. Turn the right adjuster the specified amount to obtain the proper preload. Then rotate each adjuster the same amount in opposite directions to obtain the specified backlash.

Question #108
Answer A is correct. Only Technician A is correct. A strobe light is used to identify the heavy spot on a driveshaft. Screw-type hose clamps are placed at the rear of the driveshaft with the screw heads placed opposite of the heavy spot.
Answer B is wrong. Driveshafts may be balanced in the vehicle or sent out for balancing.
Answer C is wrong. Only Technician A only is correct.
Answer D is wrong. Only Technician A is correct. Before checking driveshaft balance, always inspect the shaft for damage. A missing balance weight, accumulation of dirt, or excessive undercoating will affect the balance. To check driveshaft balance, chalk mark the driveshaft at four locations 90 degrees apart slightly in front of the driveshaft balance weight. Hold a strobe light against the rear axle housing just behind the pinion yoke. Run the vehicle in gear until the driveshaft vibration is the most severe. Point the strobe light at the chalk marks on the driveshaft and note the position of one reference mark. The number that appeared on the strobe light should have two screw type clamps installed on the shaft near the rear with the heads of the clamp opposite the number that appeared. The vehicle's suspension should have the weight of the vehicle on it when this procedure is performed so that the suspension is at its normal ride height and there are no abnormal pinion angles.

Question #109
Answer A is wrong. The driveshaft should be inspected for damage before balancing.
Answer B is wrong. Screw-type hose clamps can be used for balance weights.
Answer C is wrong. A strobe light should be used to view the chalk marks on the driveshaft.
Before checking driveshaft balance, always inspect the shaft for damage. A missing balance weight, accumulation of dirt, or excessive undercoating will affect the balance. To check driveshaft balance, chalk mark the driveshaft at four locations 90 degrees apart, slightly in front of the driveshaft balance weight. Hold a strobe light against the rear axle housing, just behind the pinion yoke. Run the vehicle in gear until the driveshaft vibration is the most severe. Point the strobe light at the chalk marks on the driveshaft and note the position of one reference mark. The number that appeared on the strobe light should have two screw type clamps installed on the shaft near the rear, with the heads of the clamp opposite the number that appeared. The vehicle's suspension should have the weight of the vehicle on it when this procedure is performed so that the suspension is at its normal ride height and there are no abnormal pinion angles.
Answer D is correct. The suspension should be supported at its normal ride height.

Question #110
Answer A is correct. Friction plates have a minimum thickness and should be measured with a micrometer.
Answer B is wrong. A feeler gauge could only measure the clearance between plates.
Answer C is wrong. Visual inspection is not accurate enough for measurement purposes.
Answer D is wrong. The friction plates must be removed from the clutch pack to measure specific plates.
The friction plates have a minimum thickness specification and should be measured with a micrometer.
The friction plate has to be removed from the clutch pack for this measurement. There is no way to measure a friction plate with a feeler gauge.

Question #111
Answer A is wrong. 10W-30 oil cannot be used on a limited slip differential.
Answer B is correct. The wrong lubricant used in a limited-slip differential will cause chatter.
Answer C is wrong. Only Technician B is correct.
Answer D is wrong. Only Technician B is correct. Limited slip differentials require a special lubricant specified by the vehicle manufacturer. The use of improper lubricant in a limited slip differential may result in differential noise, such as chattering while cornering. Clicking while turning a corner may be caused by worn limited slip components, such as clutches, preload springs, and shims. Limited slip differentials should be inspected for fiber and metal cuttings in the bottom of the housing. With the differential components removed, these cuttings may be flushed out of the housing with a solvent gun and an approved cleaning solution. Wipe the bottom of the housing out with a clean shop towel. Always use new gaskets when the differential and the cover are installed in the housing. Fill the differential to the bottom of the filler plug opening with the vehicle manufacturer's specified limited slip differential lubricant.

Question #112
Answer A is correct. The reading on the dial indicator must be subtracted from 0.100 inch to obtain the normal pinion depth shim thickness. Therefore, the nominal shim thickness is 0.043 inch and the pinion marking of -4 is subtracted from this figure.
Answer B is wrong. A 0.041 inch (1.04 mm) shim would be too thick.
Answer C is wrong. A 0.042 inch (1.07 mm) shim would be too thick.
Answer D is wrong. A 0.043 inch (1.09 mm) shim would be too thick.
The primary purpose of adjusting the pinion gear depth it to set the pinion and ring gear mesh. Pinion depth setting is the distance between a point—usually the gear end—of the pinion gear and the centerline of the axles or differential case/carrier bearing bores. The pinion gear depth is normally adjusted by installing shims onto the pinion mounting. These measurements must be taken with the pinion bearings preloaded.

Question #113
Answer A is wrong. Pinion bearing preload is measured with a dial type torque wrench on the pinion nut and rotating.
Answer B is wrong. Pinion bearing preload is measured with a dial type torque wrench on the pinion nut and rotating.
Answer C is wrong. Pinion bearing preload is measured with a dial type torque wrench on the pinion nut and rotating.
The pinion bearing preload is measured with an inch-pound torque wrench and socket installed on the pinion nut. Prior to this measurement, the pinion shaft assembly should be installed with the bearings lubricated, a new collapsible spacer, and the proper pinion depth shim. A new pinion shaft nut should be installed. Tighten the pinion nut gradually and keep measuring the turning torque. When the specified turning torque is obtained, the pinion bearing preload is correct. Never loosen the pinion nut to obtain the proper turning torque. If the pinion nut is over-tightened and the turning torque is excessive, install a new collapsible spacer and repeat the procedure.
Answer D is correct. Pinion bearing preload is measured with a dial type torque wrench on the pinion nut and rotating.

Question #114
Answer A is wrong. Technician B is also correct. Never use a torch on an axle. The axle is hardened steel and will lose its temper if heated.
Answer B is wrong. Technician A is also correct. You can remove it with a hammer but must use a stud installer to replace it.
Answer C is correct. Both Technicians are correct.
Answer D is wrong. Both Technicians are correct. When replacing a wheel stud, a hammer can be used to carefully tap the old stud out, but when installing a wheel stud, an installation tool should be used to avoid damage to the new stud. A torch should never be used to burn out a wheel stud because damage to the axle shaft and axle seal may occur.

Question #115
Answer A is wrong. The friction plates are splined to the side gear that is splined to the axle shaft.
Answer B is correct. Only the friction plates are splined to the axle shafts.
Answer C is wrong. The preload spring applies pressure to the clutch packs.
Answer D is wrong. Special lubricant is required.
Limited slip differentials have a set of multiple disc clutches behind each side gear to control differential action. The steel plates in each clutch set are splined to the case/carrier, and the friction plates between the steel placed are splined to the side gear clutch hub. Each clutch set has a preload spring that applies initial force to the clutch packs. A steel shim located in each clutch set controls preload. A friction plate always is placed next to the hub. Clicking while turning a corner may be caused by worn limited slip components, such as clutches, preload springs, and shims. Limited slip differentials should be inspected for fiber and metal cuttings in the bottom of the housing.

Question #116
Answer A is wrong. Technician B is also correct. The wheel will vibrate more as the vehicle accelerates.
Answer B is wrong. Technician A is also correct. A driveshaft will vibrate more as the vehicle accelerates.
Answer C is correct. Both Technicians are correct.
Answer D is wrong. Both Technicians are correct. Prior to driveshaft balancing, always inspect this shaft for damage, a missing balance weight, or an accumulation of dirt or undercoating. Remove both rear wheel assemblies and install the wheel nuts on the wheel studs with the flat side toward the brake drums. Chalk mark the driveshaft at four locations 90 degrees apart, just forward of the driveshaft balance weight. Fabricate a tool to hold the strobe light pickup against the rear axle housing just behind the pinion yoke. Run the vehicle in gear until the driveshaft vibration is most severe. Do not exceed 55 mph, because 55 mph indicated on the speedometer has a possible tire speed of 110 mph. Point the strobe light at the chalk marks on the driveshaft and note the position of one reference mark.
Gently apply the brakes and shut off the engine. Rotate the driveshaft until the chalk mark is in the same position as it appeared under the strobe light. Install two screw-type hose clamps on the driveshaft near the rear of the shaft. Position the clamps so both heads are beside each other and directly opposite the number that appeared under the strobe light.

Question #117

Answer A is wrong. A feeler gauge is used to measure the blocking ring to gear face clearance.

Answer B is wrong. A feeler gauge is used to measure the blocking ring to gear face clearance.

Answer C is correct. A feeler gauge is used to measure the blocking ring to gear face clearance.

Answer D is wrong. A feeler gauge is used to measure the blocking ring to gear face clearance.

The blocking ring dogteeth tips should be pointed with smooth surfaces. Clearance between the blocking ring and the matching gear dogteeth is important for proper shifting. The synchronizer sleeve must slide freely on the synchronizer hub. The threads on the blocking ring in the cone area must be sharp to get a good bite on the gear to stop it from spinning and to make a synchronized non-clashing shift. If the clearance between the blocking ring and the fourth-speed gear dogteeth is less than specified, the blocking ring is worn, which results in hard shifting. This problem would not result in noise while driving in fourth gear.

Question #118

Answer A is wrong. The ring gear runout should be checked to determine the condition of the trueness of the ring gear.

Answer B is wrong. The case/carrier side play should be measured before disassembly to determine the condition of the case/carrier.

Answer C is wrong. Bearing caps should be marked and reinstalled in the same position that they were installed.

Prior to installing the case/carrier and ring gear assembly, the side bearing and bearing cups must be lubricated with the vehicle manufacturer's recommended differential lubricant. Install the case/carrier and ring gear assembly with the bearing cups and shims or adjuster nuts. Be sure the bearing cap threads are properly seated in the adjuster nut threads when the gearing caps are installed. Tighten the bearing cap bolts to the specified torque.

Answer D is correct. The side bearings should be lubricated before installation.

Question #119

Answer A is wrong. Axle seals are not made of friction material.

Answer B is wrong. Pinion seals are not made of friction material.

Answer C is correct. Limited slip disc contains friction material.

Answer D is wrong. Gaskets are not made of friction material.

When filling a transmission or differential case/carrier, fluid will come out of the fill hole as it reaches the full level. After draining differential fluid, inspect it for excessive metal particles. Silver- or steel-colored particles are signs of gear or bearing wear. Copper- or bronze-colored particles are signs of limited slip clutch disc wear.

Question #120

Answer A is wrong. The extension housing bushing should be inspected.

Answer B is wrong. The driveshaft yoke should be inspected.

Answer C is wrong. The extension housing gasket should be inspected.

Excessive output shaft or excessive input shaft end play results in lateral shaft movement that may adversely affect the extension housing seal. A worn output shaft bearing will not cause premature extension housing seal failure. If the driveshaft yoke has a score or imperfection on the shaft, it could damage the seal and cause the transmission to leak fluid at the rear of the transmission. Replacing the seal will not correct this condition until the yoke is replaced.

Answer D is correct. The input shift bearing has no effect on the extension housing seal.

Question #121

Answer A is wrong. Technician B is also correct. A rubber gasket is used to seal components without additional sealant.

Answer B is wrong. Technician A is also correct. A cork gasket is used to seal components without additional sealant.

Answer C is correct. Both Technicians are correct.

Answer D is wrong. Both Technicians are correct. A cork gasket should be installed as it is when it comes out of the box. It is made to be installed dry and does not require any type of added sealant to help the gasket seal any better. Sometimes a spray adhesive may be used to hold the gasket in place to help installation. A rubber gasket should not use any additional gasket sealant when installed. It will become too slippery, and it may not position correctly when being installed.

Question #122
Answer A is wrong. Metallic material should be checked for during a transmission oil change.
Answer B is wrong. Type of oil and condition should be checked during a transmission oil change.
Answer C is wrong. Leaks should be checked during a transmission oil change.
During transmission or transaxle overhaul, drain and inspect the fluid. Gold-color particles in the fluid are from the wearing of the brass blocking rings on the synchronizers. Metal shavings in the fluid are produced from the wearing of the gears. An excessive amount of shavings in the fluid indicates severe gear and synchronizer wear.
Answer D is correct. Manual transaxles do not have filters.

Question #123
Answer A is wrong. The flywheel should be resurfaced every time the clutch assembly is replaced.
Answer B is correct. Only Technician B is correct. The flywheel should be resurfaced every time the clutch assembly is replaced.
Answer C is wrong. Only Technician B is correct.
Answer D is wrong. Only Technician B is correct. If too much material is removed from the flywheel, the torsion springs on the clutch plate are moved closer to the mounting bolts on the flywheel, and these springs may contact the heads of the flywheel bolts. Removing excessive material from the flywheel moves the pressure plate forward, away from the release bearing. This action increases free play; so the slave cylinder rod may not move far enough to release the clutch. By not resurfacing the flywheel, damage or premature wear can be caused to the rest of the new assembly that was installed. A flywheel should be resurfaced every time a clutch assembly is replaced. Resurfacing the flywheel will ensure that it has the flatness and a microfinish it needs so that the new clutch disc breaks in properly. If it is not resurfaced, the clutch disc will probably glaze and chatter.
The flywheel should be cleaned with hot water and soap to remove all residue on the surface after it has been resurfaced.

Question #124
Answer A is wrong. Low fluid level can cause transfer case noise.
Answer B is wrong. Transfer case chain misalignment can cause transfer case noise.
Answer C is correct. The U-joints are outside the transfer case.
Answer D is wrong. A transfer case output bearing can cause transfer case noise.
Worn U-joints may cause a squeaking or clunking noise, and a vibration while driving straight ahead.
Worn outer front drive axle joints on a 4WD vehicle may cause a vibration while cornering.

Question #125
Answer A is wrong. The vehicle speed must be below 3 mph before the transfer case will shift into 4L.
Answer B is wrong. The transmission must be in neutral before the transfer case will shift into 4L.
Answer C is wrong. The vehicle speed must be below 3 mph before the transfer case will shift into 4L, and the transmission must be in neutral.
Some 4WD vehicles have an electronically shifted transfer case that contains an electric shift motor, a magnetic clutch, a shift position sensor, and a vehicle speed sensor (VSS). This electronic shift system allows shifting between 2H and 4H at any vehicle speed. Therefore, these systems may be referred to as shift-on-the-fly systems. Shifts between 4H and 4L require the vehicle to be stopped, or moving below 3 mph, and the automatic transmission must be shifted into neutral. If the vehicle has a manual transmission, the clutch must be disengaged to allow 4H to 4L shifts. This transfer case system uses an automatic front hub lock/part-time system, which allows all front-drive components to be stationary in 2WD.
Answer D is correct. The vehicle speed must be below 3 mph before the transfer case will shift into 4L, and the transmission must be in neutral.

Glossary

Abrasion Wearing or rubbing away of a part.

Acceleration An increase in velocity or speed.

Adhesives Chemicals used to hold gaskets in place during the assembly of an engine. They also aid the gasket in maintaining a tight seal by filling in the small irregularities on the surfaces and by preventing the gasket from shifting due to engine vibration.

Alignment An adjustment to a line or to bring into a line.

Antifriction bearing A bearing designed to reduce friction. This type of bearing normally uses ball or roller inserts to reduce the friction.

Anti-seize Thread compound designed to keep threaded connections from damage due to rust or corrosion.

Automatic locking/unlocking hubs Front wheel hubs that can engage or disengage themselves from the axles automatically.

Axial Parallel to a shaft or bearing bore.

Axis The centerline of a rotating part, a symmetrical part, or a circular bore.

Axle The shaft or shafts of a machine upon which the wheels are mounted.

Axle carrier assembly A cast-iron framework that can be removed from the rear axle housing for service and adjustment of the parts.

Axle housing Designed in the removable carrier or integral carrier types to house the drive pinion, ring gear, differential, and axle shaft assemblies.

Axle shaft A shaft on which the road wheels are mounted.

Axle-shaft end thrust A force exerted on the end of an axle shaft that is most pronounced when the vehicle turns corners and curves.

Axle-shaft tube A tube attached to the axle housing center section to surround the axle shaft and bearings.

Backlash The amount of clearance or play between two meshed gears.

Balance Having equal weight distribution. The term is usually used to describe the weight distribution around the circumference and between the front and back sides of a wheel and tire assembly.

Ball bearing An antifriction bearing that consists of a hardened inner and outer race with hardened steel balls that roll between the two races and support the load of the shaft.

Ball-and-trunnion universal joint A non-constant-velocity universal joint that combines the universal joint and slip joint.

Ball joint A suspension component that attaches the control arm to the steering knuckle and serves as the lower pivot point for the steering knuckle.

Bearing The supporting part that reduces friction between a stationary and rotating part or between two moving parts.

Bearing caps In the differential, caps held in place by bolts or nuts that, in turn, hold bearings in place.

Bearing cone The inner race, rollers, and cage assembly of a tapered-roller bearing which must always be replaced in matched sets.

Bearing cup The outer race of a tapered-roller bearing or ball bearing.

Bearing race The surface on which the rollers or balls of a bearing rotate. The outer race is the same thing as the cup, and the inner race is the one closest to the axle shaft.

Belleville spring A tempered spring steel cone-shaped plate used to aid the mechanical force in a pressure plate assembly.

Bell housing A housing that fits over the clutch components and connects the engine and the transmission.

Bellows Rubber protective covers with accordion-like pleats used to contain lubricants and keep out contaminating dirt or water.

Bevel spur gear Gear that has teeth with a straight centerline cut on a cone.

Boot A term used for bellows.

Brinnelling Rough lines worn across a bearing race or shaft due to impact loading, vibration, or inadequate lubrication.

Burr A feather edge of metal left on a part being cut with a file or other cutting tool.

Bushing A cylindrical lining used as a bearing assembly; made of steel, brass, bronze, nylon, or plastic.

C-clip A C-shaped clip used to retain the drive axles in some rear axle assemblies.

Canceling angles Opposing operating angles of two universal joints cancel the vibrations developed by the individual universal joint.

Cardan universal joint A non-constant-velocity universal joint consisting of two yokes with their forked ends joined by a cross. The driven yoke changes speed twice in 360 degrees of rotation.

Castellated nut A nut with six raised portions or notches through which a cotter pin can be inserted to secure the nut.

Center hanger bearing Ball-type bearing mounted on a vehicle crossmember to support the driveshaft and provide better installation angle to the rear axle.

Center section The middle of the integral axle housing containing the drive pinion, ring gear, and differential assembly.

Centering joint Ball socket joint placed between two Cardan universal joints to ensure that the assembly rotates on center.

Centrifugal clutch A clutch that uses centrifugal force to apply a higher force against the friction disc as the clutch spins faster.

Chamfer A bevel or taper at the edge of a hole or a gear tooth.

Chamfer face A beveled surface on a shaft or part that allows for easier assembly.

Chase To straighten up or repair damaged threads.

Chassis The vehicle frame, suspension, and running gear.

Chuckle A rattling noise that sounds much like a stick rubbing against the spokes of a bicycle wheel.

Circlip A split steel snap ring that fits into a groove to hold various parts in place.

Clamp bolt Another name for a pinch bolt.

Clashing Grinding sound heard when gear and shaft speeds are not the same during a gearshift operation.

Clearance The space allowed between two parts, such as between a journal and a bearing.

Close ratio A relative term for describing the gear ratios in a transmission.

Clunking A metallic noise most often heard when a transmission is engaged in reverse or drive, caused by excessive backlash somewhere in the driveline and felt or heard in the axle.

Cluster assembly A manual transmission related term applied to a group of gears of different sizes machined from one steel casting.

Clutch A device for connecting and disconnecting the engine from the transmission or for a similar purpose in other units.

Clutch control cable A cable assembly with a flexible outer housing anchored at the upper and lower ends with an inside cable that transfers clutch pedal movement to the clutch release lever.

Clutch (friction) disc The friction material part of the clutch assembly that fits between the flywheel and pressure plate.

Clutch fork A Y-shaped member into which the throwout bearing is assembled.

Clutch housing A large aluminum or iron casting that surrounds the clutch assembly, located between the engine and transmission.

Clutch linkage A combination of shafts, levers, or cables that transmits clutch pedal motion to the clutch assembly.

Clutch packs A series of clutch discs and plates installed alternately in a housing to act as a driving or driven unit.

Clutch pedal A pedal in the driver's compartment that operates the clutch.

Clutch push rod A solid or hollow rod that transfers linear motion between movable parts; that is, the clutch release bearing and release plate.

Clutch shaft A term used for the transmission input shaft or main drive pinion. The clutch driven disc drives this shaft.

Clutch slippage A condition whereby engine speed increases but increased torque is not transferred through to the driving wheels.

Clutch teeth The locking teeth of a gear.

Coil spring A heavy wire-like steel coil used to support the vehicle weight while allowing for suspension motions.

Coil preload springs Coil springs that are made of tempered steel rods formed into a spiral that resist compression; located in the pressure plate assembly.

Coil spring clutch A clutch using coil springs to hold the pressure plate against the friction disc.

Companion flange A mounting flange that attaches a driveshaft to another drive train component.

Cone clutch The driving and driven parts conically shaped to connect and disconnect power flow. A clutch made from two cones, one fitting inside the other. Friction between the cones forces them to rotate together.

Constant mesh Manual transmission design permits the gears to be constantly enmeshed regardless of vehicle operating circumstances.

Constant mesh transmission A transmission in which the gears are engaged at all times, and shifts are made by sliding collars, clutches, or other means to connect the gears to the output shaft.

Constant-velocity joint (CV joint) A flexible coupling between two shafts that permits each shaft to maintain the same driving or driven speed regardless of operating angle, allowing for a smooth transfer of power.

Cotter pin A type of fastener, made from soft steel in the form of a split pin, that can be inserted in a drilled hole. The split ends are spread to lock the pin in position.

Counterclockwise (ccw) rotation Rotating the opposite direction of the hands on a clock.

Countergear assembly A cluster of gears designed on one casting with short shafts supported by antifriction bearings.

Countershaft An intermediate shaft that receives motion from a mainshaft and transmits it to a working part, sometimes called a lay shaft.

Coupling A connecting means for transferring movement from one part to another; may be mechanical, hydraulic, or electrical.

Coupling yoke A part of the double Cardan universal joint that connects the two universal joint assemblies.

Cover plate A stamped steel cover bolted over the service access to the manual transmission.

Crossmember A steel part of the frame structure that transverses the vehicle body to connect the longitudinal frame rails. Crossmembers can be welded into place or removed from the vehicle.

CV joint Constant-velocity joint.

Dead axle An axle that only supports the vehicle and does not transmit power.

Detent A small depression in a shaft, rail, or rod into which a pawl or ball drops when the shaft, rail, or rod is moved, providing a locking effect.

Dial indicator A measuring instrument with the readings indicated on a dial rather than on a thimble as on a micrometer.

Diaphragm spring A circular disc, shaped like a cone, with spring tension that allows it to flex forward or backward.

Diaphragm spring clutch A clutch in which a diaphragm spring, rather than a coil spring, applies pressure against the friction disc.

Differential A mechanism between drive axles that permits one wheel to run at a different speed than the other while turning.

Differential action An operational situation where one driving wheel rotates at a slower speed than the opposite driving wheel.

Differential case The metal unit that encases the differential side gears and pinion gears, and to which the ring gear is attached.

Differential case spread Another name for preload.

Differential drive gear A large circular helical gear driven by the transaxle pinion gear and shaft and which drives the differential assembly.

Differential housing Cast-iron assembly that houses the differential unit and the drive axles. This is also called the rear axle housing.

Differential pinion gears Small beveled gears located on the differential pinion shaft.

Differential pinion shaft A short shaft locked to the differential case. This shaft supports the differential pinion gears.

Differential ring gear A large circular hypoid-type gear enmeshed with the hypoid drive pinion gear.

Differential side gears The gears inside the differential case that are internally splined to the axle shafts, and are driven by the differential pinion gears.

Direct drive One turn of the input driving member compared to one complete turn of the driven member, such as when there is direct engagement between the engine and driveshaft where the engine crankshaft and the driveshaft turn at the same rpm.

Disengage When the operator moves the clutch pedal toward the floor to disconnect the driven clutch disc from the driving flywheel and pressure plate assembly.

Dog tooth A series of gear teeth that are part of the dog clutching action in a transmission synchronizer operation. The locking teeth of a gear.

Double Cardan universal joint A near constant-velocity universal joint that consists of two Cardan universal joints connected by a coupling yoke.

Double-offset constant-velocity joint Another name for the type of plunging, inner CV joint found on many GM, Ford, and Japanese FWD cars.

Double reduction axle A drive axle construction in which two sets of reduction gears are used for extreme reduction of the gear ratio.

Dowel pin A pin inserted in matching holes in two parts to maintain those parts in fixed relation to one another.

Downshift To shift a transmission into a lower gear.

Drive line The universal joints, drive shaft, and other parts connecting the transmission with the driving axles.

Drive-line torque Relates to rear wheel driveline and is the transfer of torque between the transmission and the driving axle assembly.

Driveline wrapup A condition where axles, gears, U-joints, and other components can bind or fail if the 4WD mode is used on pavement where 2WD is more suitable.

Drive pinion The gear that takes its power directly from the drive shaft or transmission and drives the ring gear.

Drive pinion flange A rim used to connect the rear of the drive shaft to the rear axle drive pinion.

Drive pinion gear One of the two main driving gears located within the transaxle or rear driving axle housing. Together the two gears multiply engine torque.

Drive shaft An assembly of one or two universal joints connected to a shaft or tube; used to transmit power from the transmission to the differential. Also called the propeller shaft.

Drive shaft installation angle The angle the drive shaft is mounted off the true horizontal line; measured in degrees.

Driven disc The part of the clutch assembly that receives driving motion from the flywheel and pressure plate assemblies.

Driven gear The gear meshed directly with the driving gear to provide torque multiplication, reduction, or a change of direction.

Driving axle A term related collectively to the rear driving axle assembly where the drive pinion, ring gear, and differential assembly are located within the driving axle housing.

Dry-disc clutch A clutch in which the friction faces of the friction disc are dry, as opposed to a wet-disc clutch, which runs submerged in oil. The conventional type of automobile clutch.

Dual-reduction axle A drive axle construction with two sets of pinions and gears, either of which can be used.

Dummy shaft A shaft, shorter than the countershaft, used during disassembly and reassembly in place of the countershaft.

Dynamic balance The balance of an object when it is in motion; for example, the dynamic balance of a rotating drive shaft.

Eccentric One circle within another circle wherein both circles do not have the same center or one circle is mounted off center.

Eccentric washer A normal looking washer with its hole not in its center. The hole is offset from the center.

Emulsification When water droplets mix with grease resulting in a thicker solution than normal grease.

End play The amount of axial or end-to-end movement in a shaft due to clearance in the bearings.

Engage When the vehicle operator moves the clutch pedal up from the floor, this engages the driving flywheel and pressure plate to rotate and drive the driven disc.

Engagement chatter A shaking, shuddering action that takes place as the driven disc makes contact with the driving members. Chatter is caused by a rapid grip and slip action.

Engine The source of power for most vehicles. It converts burned fuel energy into mechanical force.

Extension housing An aluminum or iron casting of various lengths that encloses the transmission output shaft and supporting bearings.

External cone clutch The external surface of one part has a tapered surface to mate with an internally tapered surface to form a cone clutch.

External gear A gear with teeth across the outside surface.

Externally tabbed clutch plates Clutch plates are designed with tabs around the outside periphery to fit into grooves in a housing or drum.

Extreme pressure lubricant A special lubricant for use in hypoid gear differentials; needed because of the heavy wiping loads imposed on the gear teeth.

Final drive gears Main driving gears located in the axle area of the transaxle housing.

Final drive ratio The ratio between the drive pinion and ring gear.

First gear A small diameter driving helical- or spur-type gear located on the cluster gear assembly. First gear provides torque multiplication to get the vehicle moving.

Fit The contact between two machined surfaces.

Fixed type constant-velocity joint A joint that cannot telescope or plunge to compensate for suspension travel. Fixed joints are always found on the outer ends of the drive shafts of FWD cars. A fixed joint may be of either Rzeppa or tripod type.

Flange yoke The part of the rear universal joint attached to the drive pinion.

Fluid coupling A device in the power train consisting of two rotating members; transmits power from the engine, through a fluid, to the transmission.

Fluid drive A drive in which there is no mechanical connection between the input and output shafts and power is transmitted by moving oil.

Flywheel A heavy metal wheel that is attached to the crankshaft and rotates with it; helps smooth out the power surges from the engine power strokes; also serves as part of the clutch and engine-cranking system.

Flywheel ring gear A gear, fitted around the flywheel, that is engaged by teeth on the starting-motor drive to crank the engine.

Forward coast side The side of the ring gear tooth the drive pinion contacts when the vehicle is decelerating.

Forward drive side The side of the ring gear tooth that the drive pinion contacts when accelerating or on the drive.

Four-wheel drive (4WD) A vehicle system with driving axles at both front and rear, so that all four wheels can be driven. 4WD is the standard abbreviation for four-wheel drive.

Four wheel high A transfer case shift position where both front and rear drive shafts receive power and rotate at the speed of the transmission output shaft.

Frame The main understructure of the vehicle to which everything else is attached. Most FWD cars have only a subframe for the front suspension and drivetrain. The body serves as the frame for the rear suspension.

Freewheeling clutch A mechanical device that will engage the driving member to impart motion to a driven member in one direction but not the other. Also known as an overrunning clutch.

Friction bearing A bearing in which there is sliding contact between the moving surfaces. Sleeve bearings, such as those used in connecting rods, are friction bearings.

Friction disc In the clutch, a flat disc faced on both sides with friction material and splined to the clutch shaft. It is positioned between the clutch pressure plate and the engine flywheel. Also called the clutch disc or driven disc.

Friction facings A hard-molded or woven asbestos or paper material that is riveted or bonded to the clutch driven disc.

Front bearing retainer An iron or aluminum circular casting fastened to the front of a transmission housing to retain the front transmission bearing assembly.

Front differential/axle assembly Like a conventional rear axle but having steerable wheels.

Front-wheel drive(FWD) The vehicle has all drivetrain components located at the front.

Fulcrum rings A circular ring over which the pressure plate diaphragm spring pivots.

Full-floating rear axle An axle that only transmits driving force to the rear wheels. The weight of the vehicle (including payload) is supported by the axle housing.

Fully synchronized In a manual transmission, the synchronizer assembly operates to improve the shift quality in all forward gears.

Galling Wear caused by metal-to-metal contact in the absence of adequate lubrication. Metal is transferred from one surface to the other, leaving behind a pitted or scaled appearance.

Gasket A layer of material, usually made of cork, paper, plastic, composition, metal, or a combination of these, placed between two parts to make a tight seal.

Gasket cement A liquid adhesive material, or sealer, used to install gaskets.

Gear A wheel with external or internal teeth that serves to transmit or change motion.

Gear clash The noise that results when two gears are traveling at different speeds and are forced together.

Gear lubricant A type of grease or oil blended especially to lubricate gears.

Gear noise The howling or whining of the ring gear and pinion due to an improperly set gear pattern, gear damage, or improper bearing preload.

Gear ratio The number of revolutions of a driving gear required to turn a driven gear through one complete revolution. For a pair of gears, the ratio is found by dividing the number of teeth on the driven gear by the number of teeth on the driving gear.

Gear rattle A repetitive metallic impact or rapping noise that occurs when the vehicle is lugging in gear.

Gear reduction When a small gear drives a large gear, there is an output speed reduction and a torque increase that results in a gear reduction.

Gear whine A high-pitched sound developed by some types of meshing gears.

Gearshift A linkage-type mechanism by which the gears in an automobile transmission are engaged and disengaged.

Half shift Either of the two drive shafts that connect the transaxle to the wheel hubs in FWD vehicles, may be of solid or tubular steel and may be of different lengths.

Harshness A bumpy ride caused by a stiff suspension. Can often be cured by installing softer springs or shock absorbers.

Helical gear Gears with the teeth cut at an angle to the axis of the gear.

Herringbone gear A pair of helical gears designed to operate together. The angle of the pair of gears forms a V.

High pedal A clutch pedal that has an excessive amount of pedal travel.

Hot spots The small areas on a friction surface that are a different color, normally blue, or are harder than the rest of the surface.

Hotchkiss drive A type of rear suspension in which leaf springs absorb the rear axle housing torque.

Hub The center part of a wheel, to which the wheel is attached.

Hydraulic clutch A clutch that is actuated by hydraulic pressure; used in cars and trucks when the engine is some distance from the driver's compartment so that it would be difficult to use mechanical linkages.

Hypoid gear A gear that is similar in appearance to a spiral bevel gear, but the teeth are cut so that the gears match in a position where the shaft centerlines do not meet. They are cut in a spiral form to allow the pinion to be set below the centerline of the ring gear so that the car floor can be lower.

Hypoid gear lubricant An extreme pressure lubricant designed for the severe operating conditions of hypoid gears.

ID Inside diameter.

Inboard constant-velocity joint The inner constant-velocity joint, or the one closest to the transaxle.

Inclinometer Device designed with a spirit level and graduated scale to measure the inclination of a driveline assembly.

Index To orient two parts by marking them. During reassembly, the parts are arranged so the index marks are next to each other. Used to preserve the orientation between balanced parts.

Input shaft The shaft carrying the driving gear by which the power is applied, as to the transmission.

Inserts One of several terms that could apply to the shift plates found in a synchronizer assembly.

Inserts springs Round wire springs that hold the inserts or shift plates in contact with the synchronizer sleeve, located around the synchronizer hub.

Integral housing A rear axle housing-type where the parts are serviced through an inspection cover and adjusted within and relative to the axle housing.

Interference fit A press fit; for example, if the inside diameter of a bore is 0.001 inch (0.0254 mm) smaller than the outside diameter of a shaft, the shaft must be pressed in.

Interlock mechanism A mechanism in the transmission shift linkage that prevents the selection of two gears at one time.

Intermediate bearing plate Another name for the center support plate of a transmission.

Intermediate drive shaft Located between the left and right drive shafts, it equalizes drive shaft length.

Intermediate plate A mechanism in the transmission shift linkage that prevents the selection of two gears at one time.

Internal gear A gear with teeth pointing inward, toward the hollow center of the gear.

Joint angle The angle formed by the input and output shafts of constant-velocity joints.

Journal A bearing with a hole in it for a shaft.

Key A small block inserted between a shaft and hub to prevent circumferential movement.

Keyway A groove or slot cut to permit the insertion of a key.

Knock A heavy metallic sound usually caused by a loose or worn bearing.

Knurl To indent or roughen a finished surface.

Lash The amount of free motion in a gear train, between gears, or in a mechanical assembly, such as the lash in a valve train.

Leaf spring A spring made up of a single flat steel plate or of several plates of graduated lengths assembled one on top of another; used on vehicles to absorb road shocks by bending or flexing.

Limited slip differential A differential designed so that when one wheel is slipping, a major portion of the drive torque is supplied to the wheel with the better traction; also called a non-slip differential.

Linkage Any series of rods, yokes, and levers, etc., used to transmit motion from one unit to another.

Locked differential A differential with the side and pinion gears locked together.

Locking A condition of a bearing caused by large particles of dirt that become trapped between a bearing and its race.

Locknut A second nut turned down on a holding nut to prevent loosening.

Lockplates Metal tabs bent around nuts or bolt heads.

Lock washer A type of washer that, when placed under the head of a bolt or nut, prevents the bolt or nut from working loose.

Loe speed The gearing that produces the highest torque and lowest speed of the wheels.

Lubricant Any material, usually a petroleum product such as grease or oil, that is placed between two moving parts to reduce friction.

Lug nut The nuts that fasten the wheels to the axle hub or brake rotor. Missing lug nuts should always be replaced. Overtightening can cause warpage of the brake rotor in some cases.

Lugging A term used to describe an operating condition in which the engine is operating at too low of an engine speed for the selected gear.

Matched gearset code Identification marks on two gears that indicate they are matched. They should not be mismatched with another gearset and placed into operation.

Meshing The mating, or engaging, of the teeth of two gears.

Mounts Made of rubber to insulate vibrations and noise while they support a power train part, such as engine or transmission mounts.

Multiple disc A clutch with a number of driving and driven discs as compared to a single plate clutch.

Needle bearing An antifriction bearing using a great number of long, small-diameter rollers.

Neutral In a transmission, the setting in which all gears are disengaged and the output shaft is disconnected from the drive wheels.

Neutral start switch A switch wired into the ignition switch to prevent engine cranking unless the transmission shift lever is in neutral or the clutch pedal is depressed.

Noise Any unwanted or annoying sound.

Nominal shim A shim with a designated thickness.

Nut A removable fastener used with a bolt to lock pieces together; made by threading a hole through the center of a piece of metal that has been shaped to a standard size.

O-ring A type of sealing ring, usually made of rubber or a rubberlike material. In use, the O-ring is compressed into a groove to provide the sealing action.

OD Outside diameter.

Oil seal A seal placed around a rotating shaft or other moving part to prevent leakage of oil.

One-way clutch A term used for sprag clutch.

Outboard constant-velocity joint The constant-velocity joint closest to the wheels.

Output shaft The shaft or gear that delivers the power from a device, such as a transmission.

Overcenter spring A heavy coil spring arrangement in the clutch linkage to assist the driver with disengaging the clutch and returning the clutch linkage to the full engagement position.

Overdrive Any arrangement of gearing that produces more revolutions of the driven shaft than of the driving shaft.

Overdrive ratio Identified by the decimal point indicating less than one driving input revolution compared to one output revolution of a shaft.

Overrun coupling A free-wheeling device to permit rotation in one direction but not in the other.

Overrunning clutch A device consisting of a shaft or housing linked together by rollers or sprags operating between movable and fixed races.

Pawl A lever that pivots on a shaft. When lifted it swings freely and when lowered it locates in a detent or notch to hold a mechanism stationary.

Pedal play The distance the clutch pedal and release bearing assembly move from the fully engaged position to the point where the release bearing contacts the pressure plate release levers.

Phasing Rotational position of the universal joints on the drive shaft.

Pilot bearing A small bearing, such as in the center of the flywheel end of the crankshaft, which carries the forward end of the clutch shaft.

Pilot bushing A plain bearing fitted in the end of a crankshaft. The primary purpose is to support the input shaft of the transmission.

Pilot shaft A shaft used to align parts and that is removed before final installation of the parts; a dummy shaft.

Pinion gear The smaller of two meshing gears.

Pinion carrier The mounting or bracket that retains the bearings supporting a pinion shaft.

Pivot A pin or shaft upon which another part rests or turns.

Planet carrier In a planetary gear system, the carrier or bracket in a planetary system that contains the shafts upon which the pinions or planet gears turn.

Planet gears The gears in a planetary gearset that connect the sun gear to the ring gear.

Planet pinions In a planetary gear system, the gears that mesh with, and revolve about, the sun gear; they also mesh with the ring gear.

Planet gearset A system of gearing modeled after the solar system. A pinion is surrounded by an internal ring gear and planet gears are in mesh between the ring gear and pinion around which all revolve.

Plate loading Force developed by the pressure plate assembly to hold the driven disc against the flywheel.

Plunging action Telescoping action of an inner front-wheel-drive universal joint.

Plunging constant-velocity joint Usually the inner constant-velocity joint. The joint is designed so that it can telescope slightly to compensate for suspension motions.

Power train The mechanisms that carry the power from the engine crankshaft to the drive wheels; these include the clutch, transmission, driveline, differential, and axles.

Preload A load applied to a part during assembly so as to maintain critical tolerances when the operating load is applied later.

Press fit Forcing a part into an opening that is slightly smaller than the part itself to make a solid fit.

Pressure plate That part of the clutch that exerts force against the friction disc; it is mounted on and rotates with the flywheel. A heavy steel ring pressed against the clutch disc by spring pressure.

Propeller shaft A term used for drive shaft.

Pulsation To move or beat with rhythmic impulses.

Quadrant A section of a gear. A term sometimes used to identify the shift lever selector mounted on the steering column.

Quill shaft The term used by some manufacturers to refer to the protruding hollow shaft of the transmission's front bearing retainer.

Race A channel in the inner or outer ring of an antifriction bearing in which the balls or rollers roll.

Raceway A groove or track designed into the races of a bearing or universal joint housing to guide and control the action of the balls or trunnions.

Radial The direction moving straight out from the center of a circle. Perpendicular to the shaft or bearing bore.

Radial clearance (radial displacement) Clearance within the bearing and between balls and races perpendicular to the shaft.

Radial load A force perpendicular to the axis of rotation.

Ratcheting mechanism Uses a pawl and gear arrangement to transmit motion or to lock a particular mechanism by having the pawl drop between gear teeth.

Ravigneaux Designer of a planetary gear system with small and large sun gears, long and short planetary pinions, planetary carriers, and a ring gear.

Rear axle torque The torque received and multiplied by the rear driving axle assembly.

Rear-wheel drive (RWD) A term associated with a vehicle where the engine is mounted at the front and the driving axle and driving wheels at the rear of the vehicle.

Release bearing A ball-type bearing moved by the clutch pedal linkage to contact the pressure plate release levers to either engage or disengage the driven disc with the clutch driving members.

Release levers In the clutch, levers that are moved by throwout bearing movement, causing clutch spring force to be relieved so that the clutch is disengaged, or uncoupled, from the flywheel.

Release plate Plate designed to release the clutch pressure plate's loading on the clutch driven disc.

Removable carrier housing A type of rear axle housing from which the axle carrier assembly can be removed for parts service and adjustment.

Retaining ring A removable fastener used as a shoulder to retain and position a round bearing in a hole.

Retractor clips Spring steel clips that connect the diaphragm's flexing action to the pressure plate.

Reverse idler gear In a transmission, an additional gear that must be meshed to obtain reverse gear; a gear used only in reverse that does not transmit power when the transmission is in any other position.

Ring gear A gear that surrounds or rings the sun and planet gears in a planetary system. Also the name given to the spiral bevel gear in a differential.

Roller bearing Any tube to form a gasket of any shape.

Rubber coupling Rubber-based disc used as a universal joint between the driving and driven shafts.

Run out Deviation of the specified normal travel of an object. The amount of deviation or wobble a shaft or wheel has as it rotates.

Rzeppa constant-velocity joint The name given to the ball-type constant-velocity joint (as opposed to the tripod-type constant-velocity joint). Rzeppa joints are usually the outer joints on most FWD cars. Named after its inventor, Alfred Rzeppa, a Ford engineer.

Safety stands Commonly called jack stands; used to support a vehicle when it is raised by a jack or hoist.

Score A scratch, ridge, or groove marring a finished surface.

Scuffing A type of wear in which there is a transfer of material between parts moving against each other; shows up as pits or grooves in the mating surfaces.

Seal A material, shaped around a shaft, used to close off the operating compartment of the shaft, preventing oil leakage.

Sealer A thick, tacky compound, usually spread with a brush, which may be used as a gasket or sealant to seal small openings or surface irregularities.

Seat A surface, usually machined, upon which another part rests or seats; for example, the surface upon which a valve face rests.

Self-adjusting clutch linkage Monitors clutch pedal play through a clutch control cable and ratcheting mechanism to automatically adjust clutch pedal play.

Semifloating rear axle An axle that supports the weight of the vehicle on the axle shaft in addition to transmitting driving forces to the rear wheels.

Shift forks Mechanisms attached to shift rails that fit into the synchronizer hub for change of gears.

Shift lever The lever used to change gears in a transmission. Also, the lever on the starting motor that moves the drive pinion into or out of mesh with the flywheel teeth.

Shift rails Rods placed within the transmission housing that are a part of the transmission gearshift linkage.

Shim Thin sheets used as spacers between two parts, such as the two halves of a journal bearing.

Shudder A shake or shiver movement.

Side gears Gears that are meshed with the differential pinions and splined to the axle shafts (RWD) or drive shafts (FWD).

Side thrust Longitudinal movement of two gears.

Slave cylinder Located at a lower part of the clutch housing. Receives fluid pressure from the master cylinder to engage or disengage the clutch.

Sliding fit Where sufficient clearance has been allowed between the shaft and journal to allow free-running without overheating.

Sliding yoke Slides on internal and external splines to compensate for driveline length changes.

Sliding gear transmission A transmission in which gears are moved on their shafts to change gear ratios.

Slip fit Running or sliding fit.

Slip joint In the power train, a variable-length connection that permits the drive shaft to change its effective length.

Snap ring Split spring-type ring located in an internal or external groove to retain a part.

Solid axle A rear axle design that places the final drive, axles, bearings, and hubs into one housing.

Spalling A condition of a bearing that is caused by overloading the bearing and is evident by pits on the bearings or their races.

Speed gears Driven gears located on the transmission output shaft. This term differentiates between the gears of the countergear and cluster assemblies and gears on the transmission output shaft.

Spindle The shaft on which the wheels and wheel bearings mount.

Spiral bevel gear A ring gear and pinion wherein the mating teeth are curved and placed at an angle with the pinion shaft.

Spiral gear A gear with teeth cut according to a mathematical curve on a cone. Spiral bevel gears that are not parallel have centerlines that intersect.

Spline Slot or groove cut in a shaft or bore; a splined shaft onto which a hub, wheel, gear, etc., with matching splines in its bore, is assembled so that the two must turn together.

Splined hub Several keys placed radially around the inside diameter of a circular part, such as a wheel or driven disc.

Split pin A round split spring steel tubular pin used for locking purposes; for example, locking a gear to a shaft.

Sprag clutch A member of the overrunning clutch family using a sprag to jam between the inner and outer races used for holding or driving action.

Spring A device that changes shape when it is stretched or compressed, but returns to its original shape when the force is removed; the component of the automotive suspension system that absorbs road shocks by flexing and twisting.

Spring retainer A steel plate designed to hold a coil or several coil springs in place.

Spur gear Gears cut on a cylinder with teeth that are straight and parallel to the axis.

Squeak A high-pitched noise of short duration.

Squeal A continuous high-pitched noise.

Stabilizer bar Also called a sway bar. It prevents the vehicle's body from diving into turns.

Strut assembly Refers to all the strut components, including the strut tube, shock absorber, coil spring, and upper bearing assembly.

Stub shaft A very short shaft.

Sun gear The central gear in a planetary gear system around which the rest of the gears rotate. The innermost gear of the planetary gearset.

Sway bar Also called a stabilizer bar. It prevents the vehicle's body from diving into turns.

Synchromesh transmission Transmission gearing that aids the meshing of two gears or shift collars by matching their speed before engaging them.

Synchronize To cause two events to occur at the same time; for example, to bring two gears to the same speed before they are meshed in order to prevent gear clash.

Synchronized assembly Device that uses cone clutches to bring two parts rotating at two speeds to the same speed. A synchronizer assembly operates between two gears; e.g., first and second gear, third and fourth gear.

Synchronized blocker ring Usually a brass ring that acts as a clutch and causes driving and driven units to turn at the same speed before final engagement.

Synchronized hub Center part of the synchronizer assembly that is splined to the synchronizer sleeve and transmission output shaft.

Synchronzer sleeve The sliding sleeve that fits over the complete synchronizer assembly.

Tail shaft A commonly used term for a transmission's extension housing.

Three-quarter floating axle The axle housing carries the weight of the vehicle while the bearings support the wheels on the outer ends of the axle housing tubes.

Throw-out bearing In the clutch, the bearing that can be moved inward to the release levers by clutch-pedal action to cause declutching, which disengages the engine crankshaft from the transmission.

Thrust washer A washer designed to take up end thrust and prevent excessive end play.

Tolerance A permissible variation between the two extremes of a specification or dimension.

Torque A twisting motion, usually measured in ft.-lbs. (N•m).

Torque multiplication The result of meshing a small driving gear and a large driven gear to reduce speed and increase output torque.

Torque steer An action felt in the steering wheel as the result of increased torque.

Torque tube A fixed tube over the drive shaft on some cars. It helps locate the rear axle and takes torque reaction loads from the drive axle so the drive shaft will not sense them.

Torsional springs Round, stiff coil springs placed in the driven disc to absorb the torsional disturbances between the driving flywheel and pressure plate and the driven transmission input shaft.

Total pedal travel The total amount the pedal moves from no free play to complete clutch disengagement.

Total travel Distance the clutch pedal and release bearing move from the fully engaged position until the clutch is fully disengaged.

Traction The gripping action between the tire tread and the road's surface.

Transaxle Type of construction in which the transmission and differential are combined in one unit.

Transaxle assembly A compact housing most often used in front-wheel-drive vehicles that houses the manual transmission, final drive gears, and differential assembly.

Transfer case An auxiliary transmission mounted behind the main transmission. Used to divide engine power and transfer it to both front and rear differentials, either full time or part time.

Transmission The device in the power train that provides different gear ratios between the engine and drive wheels as well as reverse.

Transmission case An aluminum or iron casting that encloses the manual transmission parts.

Transverse Power train layout in a front-wheel-drive automobile extending from side to side.

Tripod (also called tripot) A three-prong bearing that is the major component in tripod constant-velocity joints. It has three arms (or trunnions) with needle bearings and rollers that ride in the grooves or yokes of a tulip assembly.

Tripod universal joints Universal joint consisting of a hub with three arm and roller assemblies that fit inside a casting called a tulip.

Trunnion One of the projecting arms on a tripod or on the cross of a four-point universal joint. Each trunnion has a bearing surface that allows it to pivot within a joint or slide within a tulip assembly.

Tulip assembly The outer housing containing grooves or yokes in which trunnion bearings move in a tripod constant-velocity joint.

Two-disk clutch A clutch with two friction discs for additional holding power; used in heavy-duty equipment.

Two-speed rear axle A term used for a double-reduction differential.

U-bolt An iron rod with threads on both ends, bent into the shape of a U and fitted with a nut at each end.

U-joint A four-point cross connected to two U-shaped yokes that serves as a flexible coupling between shafts.

Universal joint A mechanical device that transmits rotary motion from one shaft to another shaft at varying angles.

Universal joint operating angle The difference in degrees between the drive shaft and transmission installation angles.

Unsprung weight The weight of the tires, wheels, axles, control arms, and springs.

Upshift To shift a transmission into a higher gear.

Vibration A quivering, trembling motion felt in the vehicle at different speed ranges.

Viscous friction The friction between layers of a liquid.

Wheel A disc or spokes with a hub at the center that revolves around an axle, with a rim around the outside for mounting the tire on.

Wheel offset The amount of the wheel assembly that is to the side of the wheel's mounting hub.

Wheel shimmy A side to side oscillation usually associated with tires or steering.

Worm gear A gear with teeth that resemble a thread on a bolt. It is meshed with a gear that has teeth similar to a helical tooth except that it is dished to allow more contact.

Yoke In a universal joint, the driveable torque-and-motion input and output member, attached to a shaft or tube.

Yoke bearing A U-shaped, spring-loaded bearing in the rack-and-pinion steering assembly that presses the pinion gear against the rack.

Notes

Notes

Notes

Notes

Notes